T0403122

Virality Vitality

SUNY series in Contemporary French Thought

David Pettigrew and François Raffoul, editors

Virality Vitality

Jonathan Basile

SUNY
PRESS

Published by State University of New York Press, Albany
© 2025 State University of New York
All rights reserved
Printed in the United States of America

No part of this book may be used or reproduced in any manner whatsoever without written permission. No part of this book may be stored in a retrieval system or transmitted in any form or by any means including electronic, electrostatic, magnetic tape, mechanical, photocopying, recording, or otherwise without the prior permission in writing of the publisher.

Links to third-party websites are provided as a convenience and for informational purposes only. They do not constitute an endorsement or an approval of any of the products, services, or opinions of the organization, companies, or individuals. SUNY Press bears no responsibility for the accuracy, legality, or content of a URL, the external website, or for that of subsequent websites.

EU GPSR Authorised Representative:
Logos Europe, 9 rue Nicolas Poussin, 17000, La Rochelle, France
contact@logoseurope.eu

For information, contact State University of New York Press, Albany, NY
www.sunypress.edu

Library of Congress Cataloging-in-Publication Data
Names: Basile, Jonathan, author.
Title: Virality vitality / Jonathan Basile.
Description: Albany : State University of New York Press, [2025] | Series: SUNY series in contemporary French thought | Includes bibliographical references and index.
Identifiers: LCCN 2024042451 | ISBN 9798855801842 (hardcover) | ISBN 9798855801866 (ebook)
Subjects: LCSH: Virology—History. | Virology—Philosophy. | Viruses—Social aspects. | Diseases and history.
Classification: LCC QR370 .B37 2025 | DDC 579.2—dc23/eng/20241230
LC record available at https://lccn.loc.gov/2024042451

Contents

Prodrome

Eugene Koonin, a contemporary virologist and evolutionary theorist, takes a peculiar precaution when he attempts to define the virus. The customary definition appears straightforward: "obligate intracellular parasite."[1] For particularly open-minded investigators, this expands to "obligate symbionts" (Ryan 178–79). Koonin recognizes a curious problem with these definitions, a sort of self-contradiction or auto-immunity he attempts to inoculate himself and his corpus against:

> Here we very generally define viruses as follows: *obligate intracellular parasites or symbionts that possess their own genomes encoding information required for virus reproduction and, hence, a degree of autonomy from the host genetic system, but do not encode a complete translation system or a complete membrane apparatus.*[2] This definition applies to any "truly" selfish genetic element: The key phrase here is *encoding information required for virus reproduction and, hence, possessing a degree of autonomy from the host genetic system.* Thus, regular genes [. . .] do not fit this definition, even though they may possess some selfish properties, because they encode no dedicated "devices" for their own reproduction. (*Logic* 294; original emphasis)

This book will pursue the consequences of the risk Koonin senses here, emanating from the traditional or habitual definition of virus as parasite: it would count our very genes as viruses.

To better intuit this threat or promise, life as a gift of one or several viruses, it is necessary to consider the cunning of what we typically call infection. Viral DNA or RNA, the encoder of the "information" in Koonin's definition, cannot accomplish all the stages of its replication without invading a cell — hijacking life, using us as its machine. It requires, among other things, the "translation system" mentioned by Koonin, which builds proteins from RNA. Using the cell to rapidly multiply, then destroying it to seek new hosts, is only one of the virus's infectious potentialities. It can also deposit relatively quiescent genetic material in a cell, or even integrate it into its host's genome, where it may be copied only when that cell divides — "infecting" its offspring in a manner almost indistinguishable from genetic inheritance. Reproduction as contagion, virality vitality. One mark that distinguishes these viral remnants is that they often retain genetic sequences to re-splice copies of themselves into their host's genome. This devious "device" mentioned by Koonin, the viral *mēkhanē*, is the reason that our "human" genomes possess several times more viral fossils than they do "proper" genes. These fossils often appear silent or selfish, but at times they prove essential to us and our reproduction. If it is no less possible for our human genes to be silent or expressed, beneficial, neutral, or harmful, how, then, are we to rigorously distinguish the virus from "regular" genes?

To evade this frightful consequence, Koonin has added a graft or supplement to his definition. His repeated element about repeating elements — *"encoding information required for virus reproduction and, hence, possessing a degree of autonomy from the host genetic system"* — suggests that the only way to salvage or immunify life from virality is to create this strange figure, the *autonomous parasite*. Is there such a thing? Is there anything else? Faced with the threat of virality engulfing life "itself," one can only protect it by making the viral the most vital of all.

Why is this? Why, when we attempt to define life, does it end up resembling a virus, and why, when we attempt to define the virus, does it infect life "itself?" I am here trying to view the virus, which may be integrated into a host genome, transmitted vertically by "natural reproduction," and beneficial or even essential to its host, from the perspective of its undecidability from the vital, from our genes or organelles. Nonetheless, we would be guilty of a kind of fetishism if we thought of this viral complicity as an affliction that befell life, because of its exposure to those entities we call viruses. It is not because something, apparently circumstantially, arrives from the outside that life comes to harbor the viral within it, but the iterability that comes to be known as life presupposes its haunting by a substitutability and thus adventitiousness of everything resembling its interiority. If identity depends on reproduction (for the living as for the virus), only a future that has never finished arriving will tell what is proper or essential and what dispensable or supplantable for its iterations. We are viral from before the beginning; once a thing's identity depends on repetition, it always repeats for self and other at once.

Koonin's definition is attempting to incorporate the undecidability of these relations without losing the rigor of a distinction between vitality and virality. He recognizes that any definition broad enough to encompass all the routes by which what we call the virus achieves what seem to be its own ends would threaten the purity of life itself. An "obligate intracellular parasite or symbiont" could perfectly well describe the gene. If we would still like to distinguish viral and vital replicators — (but why? We can no longer exclude the question of our ends or drives if the distinctions we cling to are groundless) — all that remains for the credulous investigator is to make the virus the most vital of all. Those very properties that made the virus something derivative and dependent — no parasite or symbiont without some life that is there already — now come to mark the virus as more autonomous than autonomy. Every independence is dependent on its dependents; it must prove its freedom or autonomy by taking leave of these accretions, in practice or in principle, and thus its very possibility remains bound to them. This is as true of the cell and of any life taken to be originary and complete in itself as it is of the virus, which now comes to prove itself as the super-autonomous, the sur-vital or sur-vival.

What is autonomy from the autonomous but dependence itself? And yet, what autonomy could be more powerful than the one that can demonstrate it has no need even of autonomy? The virus possesses, according to Koonin, a "degree of autonomy" — whether autonomy can admit of gradations or be relativistic presupposes all the questions I am posing here. It is certainly true that elements of viral "origin," even after they have been embedded in host genomes for millions of years, may retain "information required for virus reproduction." This ability to privilege their own reproduction may come at some expense to their host or play to its benefit — the two must remain in some way compatible or they would not both exist today. The displacement or deconstruction of these basic concepts does not render such provisional distinctions impossible but holds in suspense the possibility of grounding these conflictual forces in anything like a substance, nature, or essence. In short, in something like a life.

Some undecidability of virality and vitality haunts any attempt at insight into the nature of life. Operating within a regime of apparently quite different scientific and philosophical conceptuality, and attempting to define not the virus but something much closer to life itself, the parasite nonetheless rears its head in Kant's *Critique of Judgment*.[3] To define organic being, Kant takes the tree as his example, which we will see is not simply one example among others. In addition to reproduction and growth or metabolism, which allow the species and the individual to repeat itself and thus be seen as purposive, Kant gives a third criterion, which is much more difficult to circumscribe in a word:

> Third, part of the tree also produces itself inasmuch as there is a mutual dependence between the preservation of one part and that of the others. If an eye is taken from the leaf of one tree and set into the branch of another, it produces in the alien stock a plant of its own species, and so does a scion grafted onto the trunk of another tree. Hence even in one and the same tree we may regard each branch or leaf as merely set into or grafted onto it, and hence as an independent tree that only attaches itself to another one and nourishes itself parasitically. (5:371)

Centuries separate this definition from Koonin's, yet they are mirror images. In one case, the attempt to define the virus makes it more vital than life itself; in the other, the attempt to define the organic makes it into its own parasite. It is not by chance — it is essential to the organic — that it appear auto-parasitic. The following study is an attempt to draw the consequences of this auto-hetero-virality.

Viral Traces

This book is guided by the conviction that what Derrida called *deconstruction* takes place in and as the sciences of life. This means, among other things, that the goings on we attribute to substantial or essential unities — such as "the virus" or "the cell" (or the gene, the organism, the population, the environment, and so on) — are effects of an ungrounded drift that overtakes the doctrines of positive knowledge shaped by scientists, from their taxonomies and genealogies of life and its evolution to their descriptions of molecular mechanisms of heredity. As a result, the practice of reading we could say Derrida modeled — though its "method" is its openness in the face of the unanticipatable — is necessary to grapple with the aporetic texture of life and life science. In pursuing a deconstruction of the life sciences, I am building upon recent research, nicknamed *biodeconstruction* by Francesco Vitale, that has taken off from the posthumous publication of Derrida's *Life Death* seminar, which provides a model for a Derridean reading of evolutionary theory — this field continues a line of inquiry also found in the earlier work of Hans-Jörg Rheinberger, Richard Doyle, and Vicki Kirby.

Biology is frequently said, by scientists and philosophers of science today, to be undergoing a "paradigm shift" or "revolution" that is often described as the advance from a reductionist to a holistic understanding of life and its heredity, towards a model that would be adequate to its object and itself (internally, logically consistent). Moreover, continental science studies and "new materialism" have taken certain results of this transformation, its symbiotic, plastic, or networked communitarianism, as models of the *good life*, not only reifying its ontological underpinnings but simplifying its ethico-political tendencies.[4] What has been called, in various contexts, scientific progress, revision, or revolution, cannot come to a rest in a coherent system of causality, truth, or ethics — because the most basic terms of its observations or system-building, concepts such as self-reproduction and parasitism, are undecidable and unstable. Both the intractable debates that shape a given moment of scientific practice, and the transitions that supplant or absorb them, cannot possibly follow from rational, systematic, causal, or logical necessity, but are overdetermined as a result by every whim of the institutional, social, and political inscription of the practice of science, the evolution and inheritance of its language and culture, the psycho-sexual investments of its practitioners, and yet by the more encompassing alterity that holds all these determinations in suspense as well, that crosses out each term in this list as yet another artifact of chance inheritance. The unanticipatable displacements of life or its science — which will one day dispense with even these terms — undermine our most timeworn certainties.

In the following chapters, I will focus on various discordances in the history and present of virology and biology, as well as the philosophy and sociology of science, to see them as the playing out, not of coherent positions or oppositions, but of *non-oppositional* alterity. We have already seen an instance of this in

Eugene Koonin's attempt to distinguish, by means of a conceptual definition, the gene from the virus. Self-reproduction or autonomy becomes a parasite of its parasites (dependent on its dependents that it might prove its relative "independence"), while the parasitic demonstrates a "degree of autonomy" in excess of autonomy itself. This *undecidability* circumscribes every attempt at logical, conceptual, taxonomic, or causal analysis — it is not that such categorizations do not take shape nor even that they cannot maintain a provisional functionality within certain limits, but that they derive this temporary coherence from broader contexts in which they are unstable, which haunt their operations and carry them away. The demand and even the possibility of reading, a deconstructive reading such as will be set in play or found at work in the texts and machines of the life sciences, starts from here. One can only intervene in such a scene — its reading is not an empirical observation but a transformative and traductive displacement — because there is no self-identity to the characters or terms our inheritance makes available. Their apparition must leave an unthought remanence that a parasitic ear or eye can ferret out, as the site of dismembering readings or feedings, replications to come.

Certain terms and tendencies that Derrida made use of in his writing are therefore useful for following the nonlinear trajectory of scientific practice. I'll introduce here two that recur throughout this book. First: *iterability*. Both life and science are often said to depend on reproduction or reproducibility. While definitions of life are a contested field, a history that stretches from Aristotle to many of our contemporaries makes self-reproduction the defining characteristic of the living. Moreover, knowledge only demonstrates its scientific status by a similar gesture, being reproduced or reproducible and thus proving its independence of the given context of its discovery or utterance (a particular environment or laboratory, the language, culture, and subjectivity of the individual scientist, and so on). In both cases, thought responds to a demand recognized by the philosophical tradition under the name of *essence*. The essence of a thing would make it what it is and thus make it knowable as what it is (like the sun that causes life to grow and allows it to be seen). It is the possibility out of which actuality comes to a stand, within a conditional or contingent context. Then, there is a causal origin and identity of things only to the extent that they can be treated as repetitions of themselves, as individual instantiations of an essential possibility that never manifests as such. One cannot speak of the present or presence, of the here and now, or of beings or experience, until the given has been distilled into essential possibility and accidental or parasitic accretions. This basic structure, which comes to the fore in and as science and philosophy, and which haunts the living everywhere that we undergo any act of distinction or recognition, exposes all being, experience, and knowledge to what Derrida described not as essentiality or reproducibility, but as *iterability* (*Limited Inc*). If a thing depends on repetition in order to be or be known, then it has never fully or exhaustively exposed itself in or as being or truth. Its next iteration may undermine what seemed to be proper to the body or concept, the corpus, of a given reproducible, may reveal an unanticipatable monstrosity. This necessary possibility haunts every apparent individuality or idea — that the "present" or what presents itself is never simply what it is, otherwise than "being," follows from the necessity of iterability and results in the non-oppositionality or undecidability, being haunted or possessed, parasited, by what it "is" not, for which virality has already offered a figure.

More exactingly, it would be necessary to say that even the relationship of figuration, in which one individual or idea offers the schema for another, is undermined — even as it is made possible — by iterability. In a way this is just what we have seen with "the virus," that every attempt to identify it with a given individual or a rationally defined concept leaves exposed a parasitic latching point for what such imagery or conceptuality leaves unpictured or unthought. In fact, if we were to think that the sort of disturbance we are following here, haunting the bodies and texts, the corpora, of life-science, if we thought of this as a result of *the* virus, of a given entity or class of entities, something more viral than any virus, more cunning and obscure, would already be having a laugh at our expense. If anything, attempting to name something as "the virus" and *oppose* it to life or the cell or "self-reproductive" organism is only an attempt to control the subtle subterfuge of a virality that maintains a non-oppositional relationship to anything we could think of or picture as life *proper*. There is reproduction, which is to say iterability — openness to the future — only where some other or otherness lies in wait. Not in a separate and opposed site or body, but in and as the non-self-identity of any corpus. Self-identity or self-reproduction is no longer *opposed* to difference — iterability names this interweaving (a dependence of the pure principle on what it is not and what compromises it while giving it its only chance).

Second: *trace*. In its most literal form, a trace would name something present from the perspective of its constituting absence, as something inscribed makes legible the departure of the inscription. (Thus, a text implies an act of writing or printing from which it detaches and departs, and its legibility requires a projective identification with that act of authorship, one that is inevitably unstable and reverses the order of operations between inscribing and inscribed.) Nonetheless, it is necessary to think the nonoriginariness of trace without a moment of origin prior to it. If iterability is the "nature" of all life and science, if all being and thought is exposed to a future that has never finished arriving and a past it has never finished inheriting from, then there are only the flickering, present non-presences of traces. We will see an example of this (though the structure of exemplarity, of the instantiation of a pattern or type, must also be placed in question) in a moment, when we try to take the measure of displacements in the science or concept of the virus that took place after a group of scientists synthesized poliovirus. If we were to pose a question such as "what is a virus?" or even, "what is this virus here, the one we call poliovirus?" no more rigorous answer could be given than — traces. Poliovirus is not a given entity or substance, a form, figure, or concept, not its icosahedral protein shell nor some genetic sequence or code. Rather, it is anything at all only to the extent that it can leave behind any given form, that it is open to future transformations and thus is not simply equivalent to any state so far achieved in its life history or evolutionary history. (If we were to think that the evolutionary scientist or virologist would be precisely the one capable of understanding this drift, we will see that it subverts even this form of historical-genealogical and taxonomic knowledge.) As such, every effect of individual, categorical, or conceptual unity that attaches to poliovirus — word or thing — is a *trace-effect*, not an inherent or empirical given but something that must be *read*, beyond the given, in a non-programmable and unstable fashion.

Moreover, we should not think of these trace-effects as consequences of some failure or finitude of knowledge, as merely "epistemological" hindrances imposed by our fallen consciousness on a nature that would first take place without them. That is not to say that they are simply "ontological" either, but that they subvert any attempt to find a ground or simple origin for life or science in being or knowing. In order for there to be anything at all, a natural law, a material substance, a living thing or its parasites, something must repeat. What we call natural law is only the iterability of some action or reaction whose participants take their provisional identity from the repetitions of their interaction. Similarly, every act of the living that distinguishes territories as home and foreign, recognizes objects as to be ingested or avoided, recognizes predators and mates, preserves the immunity and stability of a body by metabolism, by the excision of intruders, and so on, all of this is exposed to traces or trace-structures. These differences or distinctions must be *read,* which means that they are open to every effect of virtuality, substitutability, feigning and feigned feigning, misreading, and virality, whether or not anything we would think to call consciousness shares their stage. In fact, the necessary role of this non-conscious and non-present remanence means that there is never a fully constituted consciousness — there are no conscious traces.

Deconstruction is frequently misunderstood as being fixated on philosophy, literature, language, or theory at the expense of what is most real. I turn to the life sciences here because the effects of iterability, undecidability, trace — in short, deconstruction — are not limited to the distortion that a "human" language or thought imposes on what is originally or naturally self-identical. Everywhere that Derrida returned to the question of the animal or the living, it was to puncture the self-contained metaphysical representation of supposedly human powers of language or ideality:

> For me, there is trace as soon as there is experience, that is to say referral to otherness, différance, referral to something else, etc. Therefore, everywhere there is experience, there is trace, and there is no experience without trace. Thus everything is trace, not only what I write on paper or what I record in a machine, but when I do this, such or such gesture, there is trace. [. . .] The concept of trace, I say it in a word because this would

> require long developments, has no limit, it is coextensive with the experience of the living in general. Not only of human life, but of the living in general. Animals trace, every living being traces.[5] (*Trace* 59)

Derrida extends the trace to the domain of the "living in general" here, perhaps because he is thinking of the trace as something like the necessity of reading. Trace would be there "as soon as there is experience" because for anything to leave an impression or inscription on the receptivity of the experiencer, it must be something read or interpreted. However, we can go perhaps a step further in radicalizing this structure — and this would certainly be a step in agreement with other statements of Derrida's on the trace and deconstruction ("As If" 368). We should not think of trace-structures or deconstruction as limited to what we call life or as enabled by powers or flaws of the living being's faculties, and in fact when we witness trace-effects constituting and de-constituting the borders between cell and virus, we are observing an undecidability at what many biologists claim is the limit between life and non-life. Deconstruction is not a vitalism, and iterability and trace are necessary wherever what we think of as the nature of mechanism, natural law and automaticity, repeats itself. That this "mechanism" is as open on an unanticipatable future, and a past it has never finished inheriting from, as any supposedly "vital" or "vitalist" substance — in neither case as the novel inventiveness of a self-possessed power but as the invention of the other (what overtakes any distribution of laws and matters, rather than following from them) — is the undecidability of mechanism and vitalism, life and its others. If we were to hope to distinguish these domains by contrasting non-life as mere exteriority to life as interiority, we would find that all the effects we are accustomed to attributing to perception, thought, or experience — virtuality, substitutability, fallibility, and so on — are already there whenever two "material" things interact according to "natural law." What they are and what causes them to be can only befall them from the other (the still-to-come), which leaves them haunted by non-presence or trace-effects in the "here and now."

It is often said, in the theoretical humanities today, that the generation of thinkers including Derrida was too focused on things human, among which language and thought are hastily classed.[6] And thus, that they failed to prepare us for the increasingly urgent task of recognizing the agency of "matter" or "nature." One could turn to the texts of Derrida to recognize the untenability of such representations of his thought, which punctures such limits with every word (Fritsch et al., "Introduction"; Clark and Lynes). I turn here rather to virology and the life sciences, to discourses that Derrida rarely explicitly engaged with, for at least two reasons.[7] First, because the test of what we sometimes call "theory" is not its having-said-everything-already, nor the expectation that we will find our own concerns reflected there word-for-word, but the promise it holds for the unthought, how it opens on what it could not possibly have anticipated, certainly without remaining unchanged in the process. Second, because the natural sciences that are increasingly placed on a pedestal by humanistic discourse today (and by many others in the university and beyond), are not immune to the effects I have described here. When one avoids the pre-critical exuberance of contemporary "materialism" or "realism" — the hypostatization of "material" things, the aggrandizing of the authority of scientific or simply common sense understandings of them, the complacency of expecting a unified ethico-political tendency to emanate from them — when one approaches matter, nature, or natural science in and as deconstruction, one is not turning away from some originary wellspring of natural truth and goodness, but rather approaching what we may, for now, persist in calling nature, matter, or life, but which has never been *opposed* to trace or inscription, to an unanticipatable alterity of which "we" are merely the breeding ground.

Digest

The following introduction examines a controversial 2002 experiment, which synthesized infectious poliovirus from mail-order DNA fragments, to argue that it demonstrates the interweaving of deconstruction-effects and virality. This experiment confronted scientists, lawmakers, and society at large with questions that might previously have seemed idle and inconsequential: Is a virus matter or information? Is it fundamentally or essentially the material thing that circulates in "nature," or is it the genetic sequence that can be instantiated in a "natural template," but may be just as virulent when stored on a computer database? It is impossible in principle to answer such a question definitively — it inevitably appears at once as though things have always been this way, as if this experiment merely actualized a possibility that was always there, lying in wait, and yet even the form of the question is unthinkable without the existence of fairly recent information technologies and the scientific discourses surrounding them. Yet, despite or perhaps because of the impossibility of finding a purely theoretical or conceptual response to this question, urgent decisions seemed to follow from its unanswerability, and reconfigured every domain that suddenly recognized its exposure to an unthought viral possibility. DNA-synthesis laboratories began screening synthetic DNA orders for viral risks, everyone from virologists to lawmakers raised questions about how best to regulate this research, journalists and militaries speculated about the threats of bioterrorism it posed, and even basic definitions in the field of epidemiology — such as what it would mean for a virus to be irrevocably extinct — and the public health practices based on them, were suddenly displaced. The sympathetic resonances that set all these borders trembling are deconstruction-effects. What seems, at one and the same time, to change nothing and everything, merely by a displacement in techno-conceptual reproducibility, seizes knowledge and practice from something other than their positive face, and leaves in its wake new powers indistinguishable from impotence. In the remainder of this chapter, I contrast the theory and ethico-political claims of new materialism with Derrida's writing on viruses, to suggest a different way of understanding the responsibilities facing us in what has been called the Anthropocene, and in the midst of the COVID-19 pandemic.

The first chapter focuses on a foundational concept or figure of evolutionary science, the Tree of Life, and the displacements it is undergoing at present. The tree is a common figure in logic and natural science representing the branching off of subcategories from higher categories. In biology, it depicts taxonomic relations of descent, such as the divergence of child species from parent species. Certain phenomena whose importance is increasingly recognized today have placed the roots of this tree-structure in question. One example of a pest boring holes in biologists' trees is what they call "horizontal gene transfer." It involves the exchange of genetic material (often treated as synonymous with heredity) among only distantly related individuals, and is called "horizontal" because it moves orthogonally to the "vertical" relations of descent represented by the branching pattern of the tree. It can be pictured as horizontal lines traversing distant branches on a tree-structure, but at least among certain domains of life, it is so common that it renders fictive or impossible the very tree-structure itself. What we think of as reproduction, the filiative relationships represented by life's family tree, is a poor guide to the genetic relations of the living and to how certain traits circulate amongst them. The most cautious and reflective theorists puzzling over this predicament, including Eugene Koonin, whose work I introduced above, have recognized an irresolvable aporia in this apparent vertical-horizontal conflict. If horizontal relations are so prominent that vertical relations are mere fictions — *then there can be no "horizontal" relations either.* Without a preconception of relationships of vertical descent, the concept of horizontality has no basis. The tree is a necessarily fictive structure, but it is a *necessary* fiction — there can be no life and no science without it. Thus, all those scientists and theorists of the past several decades who have celebrated horizontality, webs, networks, or rhizomes of life, symbiosis, relationality, becoming, process, plasticity, interactionism, and so on, have merely created sects of the cult of the tree. These figures for relations across difference depend on fictive and provisional preconceptions of identity — the network or web is simply the cross section of a tree. All those who promise to put this tree to death — who draw a cross over its vertical lines — resurrect it, and this is far from the only biblical resonance attaching to the Tree of Life. The deconstructibility of life science follows from this necessary fictiveness at its foundation — one must posit a fictional narrative or structure in order to arrive even at this knowledge of its contingency, and thus every formation of knowledge by and about the living remains fragile and uprooted.

The second chapter observes this undecidability of vertical and horizontal in the instability of attempts to define the relationship of viruses to cells or organisms. One might think of verticality as proper reproduction and horizontality as contagion, but viruses demonstrate the impropriety of any such division. One means by which viruses spread seems to overtake reproduction itself and turn it into a form of contagion. The process of lysogeny involves the infection of a cell by viral genes that for the most part lie dormant in a cell, where they can be copied with the cell's inherited genes in each cycle of replication and handed down to that cell's progeny. Birth or conception "itself" is a form of infection, and I look at the past century of virology and genetics, at examples of theorists who puzzled over the similarities between genes and viruses, to read certain effects of this undecidability of germs. Next, I turn to the instability of certain phenomena perhaps related to "the virus" as an agent of horizontal transfer. One of the primary mechanisms of horizontal gene transfer among prokaryotic cells is a process known as *conjugation*, often referred to as *bacterial sex*. Conjugation allows a cell to spread some or all of its genes to a recipient, but only if the donor cell possesses a particular group of genes (the fertility factor) that the recipient lacks. The recipient receives these genes in the process, and becomes a potential donor itself, which means that the fertility factor spreads through a population like an infection, horizontally. Now, some have observed that this virus-like sex or STI may have more than a passing similarity with virality. It has been argued that the fertility factor is not "virus-like" but a "true" virus. I examine the instability of this distinction, and more fundamentally of the difference between being and seeming, as insurmountable virality-effects. While this example deals with species quite distant from our own on the "Tree of Life," I close with an example of the role "viral fossils" play in human reproduction.

The third chapter examines the resulting instability of origins and historiography by looking at intractable debates that have surrounded the "discovery" of the virus for a century now. Two oddly symmetrical debates over priority occurred in the 1890s and the 1910s, both involving competing claims to the discovery of certain phenomena now recognized as viral. At the time, there could be inklings that a novel category of organic phenomena might be implicated in the processes being observed, but it would obviously not have been possible for these theorists to declare, with anything like our understanding of the term, whether they were observing *viral* phenomena. This impossibility is more than a passing hindrance of a novel science in its nascent phase. I have already shown that today as ever this "concept" or term, virus, is in question and subject to radical displacements — it never has and never could be otherwise. There will never be an origin or discovery of anything that is not a secondary effect of this drifting uprootedness that provides temporary and contingent frameworks within which the claim of originality must be adjudicated. Every claimant seems at once confused and prescient about what they are discovering, which should not be understood as their not-yet or already knowing what it is, because we are no more secure in our own knowledge. Who discovered the virus? Well, what is that — "the virus?" These historical accounts offer a view in miniature of a generalizable tendency I am pursuing under the heading of virality, that parasitism uproots any notion of origins. In conclusion, I turn to the theories of Michel Serres and Bruno Latour, who attempt to make the history, theory, and practice of the sciences flow from the "actors themselves," for example the microbes. Their own ontotheological immanentism is displaced by the non-self-identity, the virality, of these basic categories.

The final two chapters turn toward aspects of the contemporary life sciences that do not deal explicitly with what we call viruses, in order to demonstrate that the effects of undecidability, non-oppositionality, or deconstruction that follow viruses wherever they go, nonetheless do not depend on them or any particular substance or entity to take place. They inhabit the most intimate core of anything that could be called "life itself," and disturb our attempts at scientific knowledge or technological mastery.

The fourth chapter focuses on what is called *synthetic biology* today, and in particular on a decades-long effort by scientists affiliated with the J. Craig Venter Institute to create a bacterial cell with an entirely synthetic genome. Craig Venter, a biologist and entrepreneur, has argued that this "minimal" genome project demonstrated complete knowledge of and thus complete control over the essence of life. First, I place this experiment in the context of the steady rhythm of similar claims over the past century, by biologists who have seen themselves in each case at the forefront of the technological manipulability of vital phenomena. Their claims to have exhaustively revealed the mechanisms or mechanism of life "itself," life as mere matter or a calculable engineering material, quickly grow stale as the technologies available for their models or interventions become outdated (even as their successors make identical claims, and declare their novelty in doing so). I try, then, to find a mode of writing about the present configuration of bioengineering that eschews this once-and-for-all apocalyptic fervor. The mechanism that subsumes "life" depends on a technoconceptual model that becomes outdated as quickly as our machines, which does not augur a vitalist spontaneity belonging to life itself, but overdetermines in turn every such representation of a pure power of self-invention beyond mechanism and model. What we call life is originally synthetic, which is why it remains exposed to the chance and risk of deconstruction, to being overtaken by what does not follow from any existing power or possibility, the invention of the other.

The remainder of this chapter examines the conflicts and complicities within and among the various fields (academic, commercial, legal, etc.) in which Venter and his collaborators must attempt to theorize or justify their work. In every case, it is a question of control over reproduction or reproducibility: can life be exhaustively known and manufactured, so that our power over its reproduction is total? Can we secure intellectual property for exclusive control of the reproduction of this technology? Can we control or guide the interpretation and narrative surrounding this research (its conceptual reproducibility, so to speak) so that the most benefit accrues to us? Those benefits would include the reputational, and its measures such as academic citation counts, but also the ability to *capitalize* on this research, to make capital reproduce. In every case, whether they are trying to present their results to a scientific audience, or impress a judge, an investor, or the general public, recourse to fundamental concepts is necessary. In order to present what they are doing in anything like a scientific or technological fashion — to understand it as a theory or a practice — they must draw on the conceptuality of life and its subdomains (such as species-concepts), despite the inherent instability of these concepts. It is necessary to say not just that they have created *something,* but under what conditions their act is reproducible or iterable (valid for life itself? Or only for this species, in this environment, etc.?). "Life," or some such recourse to essentiality, is necessary and impossible. It is necessary to generalize, but every generalization depends on the contexts from which it strives to break free — this infection of conceptuality by what it casts as the accidental or contingent is a form of virality or parasitism. I try to demonstrate, not merely that this exposes every scientific formation of knowledge or power to opportunistic infections, but that, as a result, everything we might think life or its science *is not* plays a role in determining its supposed interiority. The business concerns of Venter's intricate commercial empire, patent law, the professional norms of academic science and its publishing industry, the investments (in more than one sense) of venture capitalists, journalistic media and its influence over each of these domains, all play a role in shaping this research and how it gets written up even in science journals.

The fifth and final chapter looks at "de-extinction" projects, which claim to apply techniques of synthetic biology to work toward the "resurrection" of extinct species such as the woolly mammoth. Questions about scientific knowledge and its control or power over life recur here with the urgency of irreversible loss, in response to the ongoing mass extinction caused by climate change and ecological degradation. The synthetic biologists working to make "de-extinction" a reality argue that these species will restore necessary ecological functions to their former habitats, while their detractors both question the claims being made (whether this is possible and how restorative it would be) and argue that it represents a techno-optimism inimical to conservation work. At its worst, this could lead the public to think of extinction as temporary and thus undercut the urgency of conservation. In addition to typical questions about how best to preserve ecological balance, woolly mammoth de-extinction raises another question inseparable from the others: what is a mammoth? Every party to the debates for and against "de-extinction" presupposes some answer to this question, whether this sequence of technological interventions could possibly add up to something classifiable as a "woolly mammoth" and/or whether it could be relied upon as a substitute for the mammoth's ecological functions. In short, we are confronted with the question of the species-concept, notoriously a source of intractable debate among the living and our scientists of life. True knowledge of the basic concepts of life-science — such as "what is a species?" — is necessary yet impossible, one must speak as if one knew such things, and yet this very necessity exposes us to the constant chance and risk of deconstruction, of the unthought and unanticipatable carrying away what seems most confident and secure. Here, we see this stable instability, introduced by the inexhaustible demand of a question with a philosophico-conceptual and ontological form, overdetermining decisions that have the urgency of the most practical and ethical responsibilities. Decisions must be made that bear on the very future of what we call life, without our knowing what that is or could become.

This project has developed over several years, through countless unanticipated transformations, and owes its present form to everyone who has thought with me and supported my work through that time. I am most grateful for the loving support of my family, Mary Elizabeth Borkowski, Susan Basile, Edward Basile, Sarah Basile, Jacqueline Kalin, Kenneth Cooper, Michael Kalin, and Seymour. The solidarity of those I have organized with in this time has provided rare glimmers of hope without which I cannot imagine a future for anything like this work; I thank my comrades, Abby Scribner, Sarah Trebat-Leder, Dian Dian, Esra Sefik, Isaac Horwedel, Courtney Rawlings, Sarah Warren, Joe Larios, Dez Miller, Frank Voight, David Hofmann, Rebekah Spera, Sam Wohns, Liam Fox, and the current and future organizers of EmoryUnite! and the Vancouver Tenants Union. I feel that this project would not have been possible with any other group of academic advisors than those I had at Emory's Comparative Literature department: I thank Geoffrey Bennington, Elissa Marder, and Deborah Elise White. I thank as well all those faculty who provided encouragement and guidance in this project's earliest stages, including Elizabeth A. Wilson, Deboleena Roy, and Claire Nouvet. I am grateful to have done this work while surrounded by colleagues with whom I can share much more than my ideas; I thank for their friendship and insight Thomas Clément Mercier, Eszter Timár, Ryan Tracy, Francesco Vitale, Erin Graff Zivin, Peggy Kamuf, Naomi Waltham-Smith, Emile Bojesen, James Martell, Ronald Mendoza-de Jesús, Phil Lynes, Mauro Senatore, Jeremy Stewart, Erin Soros, Adam Frank, Bob Brain, Geoffrey Winthrop-Young, Alexei Kojevnikov, Brendan Moore, Matías Bascuñán, Ryan Fics, Ben Brewer, Eric Flohr-Reynolds, Natalie Catasús, and Francisco López. I thank as well Michael Rinella and the editors of SUNY Press. This research was supported by the Emory University Fox Center for Humanistic Inquiry, the University of British Columbia-Vancouver Killam Postdoctoral Research Fellowship, and the University of Toronto Arts & Science Postdoctoral Fellowship.

In loving memory of José Quiroga; I still imagine that any day one of his inimitably graceful messages will somehow reach me.

Introduction

Principles

I'll preface each chapter with a description of the scientific and theoretical concepts or terms that will be explored within it. Given that my task is demonstrating the impossibility of stable and rigorous conceptual limits, it will be difficult to provide definitions; in some cases, it will be necessary to offer a description that I will then go on to problematize.

First, the virus. *Encountered on its own (so to speak), it is composed of nucleic acids (DNA or RNA) surrounded by a protein shell and possibly an additional lipid membrane — in this phase, it is called a* virion. *There is enormous diversity in the processes by which they infect cells and replicate, but it is always the case (except under artificial conditions) that a virus cannot reproduce unless it infects a cell and makes use of the cell's machinery and resources for copying its nucleic acids and translating them into proteins, re-forming virions that can escape a cell to infect another. The order of operations most of us are taught in biology classes — DNA, transcribed into RNA, translated into proteins — is often subverted by viruses. Some store their genetic material as RNA, which can then be copied directly into another RNA, or reverse-transcribed into* cDNA *(complementary DNA — which simply refers to DNA derived from RNA), which then undergoes the normal transcription and translation processes. In what follows, it is important to note that RNA virus replication is significantly more "error-prone" than that of DNA viruses, meaning the RNA sequences copied from a parent virus will have a greater frequency of divergence.*

*In this introduction, I focus on a 2002 experiment that, from a certain perspective, transformed our understanding of viruses, though from another perspective it seems to have merely illustrated something already known. In this experiment, a team of scientists synthesized active poliovirus in their laboratory. They began, not with material derived from viruses, but with short nucleic acid molecules (*oligonucleotides*), which can be ordered by mail. They pieced these synthetic fragments together into a genetic sequence corresponding to poliovirus. This experiment caused a stir well beyond the scientific community. At first glance, it poses a relatively ontological or theoretical question: what is a virus? Is it the material structure and infective process just described, or can sequence information stored in a computer database (from which such a synthetic virus can be copied or "resurrected") be considered a real virus? On the one hand, any answer to this question is an arbitrary conceptual delimitation. On the other hand, life-or-death decisions faced by virologists, epidemiologists,* *politicians, the military, journalists, and many others, depend on the space of undecidability that opens within this question. The reconfiguration of all these institutions in the wake of this experiment allows a glimpse of the scientific process in and as* deconstruction. *I will not offer a definition of deconstruction, because it is not a concept, but what is taking place here makes legible certain deconstruction-effects, how the most urgent decisions and transformations of institutional authority follow precisely from what subverts our delineation of concepts and the disciplinary silos based on them.*

***

In 2002, three scientists, from the fields of virology and biochemistry, performed an experiment that tested the apparent limits of virality and vitality. They reproduced poliovirus from a genetic code stored not in a nucleic acid template, but in a database. They ordered synthetic DNA by mail, exposed it to transcription and translation machinery derived from human cells, and confirmed that their synthetic product was able to create "wild-type" viruses (Cello et al.).[1] Two of the authors of the original study later referred to it as "the first synthesis of a replicating organism in the absence of natural template" (Wimmer and Paul 590).[2] The experiment caused considerable controversy, and raised a series of questions that cannot be easily delimited, but certainly cross or contort the boundaries of ontology, epistemology, and ethics; science, philosophy, and politics; nature and art or technology; production and reproduction, life and death, chemical and biological, information and material, actuality and possibility — virality is the betrayal of these borders, the infection of each term with its opposite.

For instance, one could ask whether what these scientists made was poliovirus, and what that thing we call poliovirus must be if it can be created in this way. Eckard Wimmer, one of the scientists who performed the initial experiment, later described it as a displacement of the essence of virality:

> Our experiment has thus overthrown one axiom in biology — namely, that the proliferation of cells or, for that matter, viruses depends on the physical presence of a functional genome to instruct the replication process. It was believed that without parental genomes, no daughter cells or progeny viruses would arise. We have broken this fundamental law of biology by reducing poliovirus to a chemical entity, *which can be synthesized on the basis of information stored in the public domain* — an experimental proof of principle that is applicable to the synthesis of all viruses.
>
> Indeed, any DNA molecule of any length could be assembled from oligonucleotides; thus, a virus might be resurrected from pure genomic information. (S3; emphasis added)

Rudolf Virchow once declared, dismissing the notion of spontaneous generation, *omnis cellula e cellula*, which Eugene Koonin, much later, adapted for virology: *omnis virus e virus* (*Logic* 318). Every cell from cell, every virus from virus. On the one hand, Wimmer seems to be transgressing this vital principle — he has reproduced a virus, not from an *actual* or *natural* virus, (nor from "parent" to "daughter" — the computer is estranged from this family scene), but from its genetic information stored in a database. On the other hand, if one thinks of this informational pattern as a virus itself, or perhaps as *the virus itself*, as the true form or essence of virality, then the tautology is restored. A virus has been copied from a virus, just as it always has been.[3]

My intention is neither to declare that one fine day in 2002, the essence of virality shifted from "organisms circulating in nature" to their genetic formula in a data bank, nor to claim, somewhat ahistorically, that viruses have always been information. Rather, their apparent displacement is not a historical or essential fact but an effect of what we could call, in Derridean terms, *the non-self-identity of the trace*. This could be glossed as follows: the virus is no more natural than cultural or technical, no more biological than chemical, no more material than informational, and any other apparently conceptual or physical division we might attempt to institute to capture or quarantine its essence would prove similarly deconstructible. The virus, if it is anything at all, is what repeats — in order for this something here to be a virus, it must be one instance of an iterable family or structure. It is this iterability that philosophers and scientists have long sought to harness by defining life as *reproduction*. As a result, neither virus nor anything else is ever simply *this thing here*; nothing is ever but the token of a type or pattern that remains in abeyance no matter how many instantiations it seems to bring forth. The trace is no more this unknowable typology or archetype than it is the unnamable singularity which can only exist as a failure to present the law — the trace only comes to pass as this impossibility of substance and law, without which there would be nothing and yet which promises that nothing will ever simply be. Thus, "the virus" is no more a pattern imprinted in organic macromolecules than it is the substitute pattern in computerized memory. Both sites are stand-ins or traces that are never simply present and have no proper abode but leave a haunting remainder or *restance* no matter what form or in-formation they seem to take. Never having a body of its own, occupying and perverting a borrowed site — in this the trace resembles a parasite, a virus.

Indeed, this experiment and Wimmer's interpretation of it reveal more than anything the contingent and contextual form that any ontological question must take. Rather than offering an ultimate, timeless, or eternal answer to a question such as "what is a virus?," Wimmer demonstrates that even posing such a question requires terms and technologies whose provisional and contingent nature overdetermines any ontology, philosophy, or science. For any number of reasons, a question such as "is a viral genetic sequence stored in a computer database a real virus?" would have been almost unthinkable a century ago. In addition to the particular forms of information technology on which the question depends, an entire doctrine of knowledge about virology and the material basis of genetics or natural heredity is necessary, as well as a vast technological apparatus of genetic manipulation or bioengineering. This technics is anything but supernumerary to that doctrine of apparently natural or biological genetics — the manipulability (theoretical and practical) of viral and vital reproductions determines the forms scientific knowledge of its causes can take. Like any supplement, these technics are essential, or one might say vital, to the very thing they seem to imitate or attach to extrinsically. Similarly, neither genetic science nor genetic engineering reached their current form without a theory of information and the cybernetic program, which became the dominant "model" deployed to understand the genetic "program" (Mayr, "Cause"; Jacob; Keller, *Making Sense*). Certain technologies of reproducibility, in the given form of their present familiarity, are necessary to answer or just to frame any question of essence, even that of the "natural" essence of the living (already for Plato, defining the true Forms required exiling shadows, reflections, paintings, and so on). It is not possible to say that such an experiment has revealed what the virus has always been, nor can we know that its nature has simply been transformed by this experiment. The one thing that always "self"-reproduces is this limit, that only allows for knowledge or being to take a contingent and surmountable form, whether that is understood as boundaries (or enabling constraints) imposed by technology, a doctrine or paradigm of scientific interpretation, the arbitrariness of natural language, or other "cultural" constraints.

Neither revision nor revolution. Even if the regime of nature and technics that makes synthetic virology and biology possible is quite young, it is not impossible to tie it back to something that has been imaginable throughout history, in the form of the golem, the automaton, Pygmalion's statue, Frankenstein's monster, and so on. It is never simply the case that one invents, making the previously impossible possible, nor that one discovers how to make what has always been possible actual — there are only *translations* of one set of apparent limits into another, which can never promise that they will have been faithful. Every translation translates the *untranslatable*, which is neither a necessity, given that it can only be expressed within a historical regime of language, conceptuality, and technology, nor a contingency, given that there is nothing at all without such limits.

One might think that the virus is something much older than the computer, but their definition as objects of inquiry is almost contemporaneous (and their foreshadowings can be found no matter when or where one looks). By necessity, for reasons that are not at all specific to the virus or its field, there is no definitive answer as to when the virus enters science. If one examines the field completely naïvely and without any attempt at contextualization, one can identify a time when "virus" is an object of study for those calling themselves "virologists." But how to date the origin of *what we call* virus is much trickier, if we confront the fact that the mere occurrence of the word is no guarantee of that essential sameness. The problem is especially bottomless if we acknowledge that *we do not know what that is* — no doubt it will seem different tomorrow or the day after. It has never been contemporaneous with itself. There were certainly those who called themselves "virologists" before there was a cybernetic theory of the genetic program, but if that theory displaced what we thought we knew of life and of its viruses, then it is not so easy to dismiss the idea that "the virus" has always been a computer or something computational. One could repeat these analyses with respect to computation or information, which have never simply been absent or present.

We will see, in a moment, that however "theoretical" or abstract these questions and their unanswerability may seem, they are co-implicated in decisions of the most "practical" or urgently ethico-political sort, demanding a reckoning from or shifting the ground under academic and commercial scientific laboratories, epidemiologists, journalists, politicians, and defense departments. The countless forms of power and knowledge implicated in these domains find their fundamental commonality not in a shared essence from which they derive, but in their mutual exposure to viral subversion, to the sway of what is not and has no proper form.

It is necessary to distinguish the deconstruction or displacement that is underway here, whose course I am merely following, from those tendencies in "materialist" thought today to celebrate the crossing of borders. It is not *the* virus that entangles nature and culture or the material and semiotic, because such a thing can only appear to complicate these borders if they have already been set in place by decisions that are neither innocent nor absolute, and whose deconstruction displaces any supposed positivity or agency of the virus, its matter, and everything else. Thus, the movement of virality is not assimilable to those hybrid figures that are hypostasized or fetishized when taken as emblematic of the "agency" of "matter:" symbioses and the wood wide web, the heterogeneity of Latourian networks, or the rhizomes and becomings of Deleuzian biophilosophy. The risk of celebrating such border-crossings is always that one will reify and naturalize the underlying borders in the process. We can recognize this pattern in the work of Donna Haraway, for instance, in the paragons of "natureculture" and "multi-species coflourishing" she elects, figures such as symbiosis and sympoiesis, companion species, transgenic organisms, and the cyborg (*Staying*; *When*; *Modest_Witness*; "Cyborg"). It is only possible to cross borders if they have been posited in advance — thus, every exhortation today to see nature and culture entangled or without separation is simultaneously, in the same breath, an invocation of separation and pure origins. This is quite apparent in Haraway's tendency to celebrate the crossing of genetic lineages — for instance, a genetically engineered mouse designed to develop cancer and thus allow for study of the disease, the Oncomouse, becomes one figure for transformative kinship: "Transgressive border-crossing pollutes lineages — in a transgenic organism's case, the lineage of nature itself — transforming nature into its binary opposite, culture" (*Modest_Witness* 60). It is not possible to identify the transgenic organism with a transgression of "the lineage of nature itself" unless one has accepted the most dogmatic Modern Synthesis version of genetic determinism guiding the heredity of life. Without the prior acceptance of some such model, there is no fixed measure for what borders are or are not crossed, how and to what degree, by any apparent filiation or contagion. Deconstruction or virality is not a something that could be *opposed* to such borders or their artificiality; the only chance and risk of an event, such as Wimmer's experiment perhaps was, depends on its inscription in a conceptual structure or family scene. The virality of borders is just as much their crossing and contamination as it is their institution and maintenance — it is a corruption and complication without which the simplicity of origin could not exist.

In the terms deployed by Haraway, we could ask: Can there be the sym- of sym-biosis or sym-poiesis without the auto-? Difference without identity? Is there any sense in *opposing* these two, as if they were separate and separable logics and models, between which we could freely choose? For instance, when Haraway writes, "Recuperation is still possible, but only in multispecies alliance, across the *killing* divisions of nature, culture, and technology and organism, language and machine" (*Staying* 117–18; emphasis added), is there any sense in claiming that divisions *kill* if it is only on their basis that the multi-ness of "multispecies alliance" can be thought? That is to say, if difference, division, and its very artificiality is what *gives life?*

Quasi-Species

Wimmer's contextualization of this experiment depends on drawing a stark distinction between the chemical and biological, though these fields' shared dependence on trace-structures renders their opposition quite fragile. By his framing, only the chemical nature of a virus is captured by a computer, while it retains a biological nature that surpasses informatization. He always places the computerized viral code on the side of the chemical — for example, "For many virologists, the dual nature of viruses as chemicals with formulae stored in data banks and as organisms circulating in nature was not news" (Wimmer S9). To emphasize this distinction, he makes a point of writing out the chemical formula of poliovirus: $C_{332,652}H_{492,388}N_{98,245}O_{131,196}P_{7,501}S_{2,340}$. It is not irrelevant to note that *formula* is also one translation of Ancient Greek *logos*, and that the trace disturbs any pretense of *logocentrism* — the necessary but impossible abstractability of any term or germ into a purely reproducible ideality.

Wimmer's attempt to maintain this distinction between the chemical and biological requires raising the question of the nature or definition of life, in the classical form of a tension between mechanism and vitalism. The virus crosses the divide of this apparent opposition:

> When I am asked whether poliovirus is a non-living or a living entity, my answer is yes. I regard viruses as entities that alternate between non-living and living phases. Outside the host cell, poliovirus is as dead as a ping-pong ball. It is a chemical that has been purified to homogeneity and crystallized, with its physical and chemical properties largely determined, and its three-dimensional structure solved. Just like a common chemical, poliovirus has been synthesized in the test tube.
>
> Once poliovirus, the chemical, has entered the cell, however, it has a plan for survival. Its proliferation is then subject to evolutionary laws: heredity, genetic variation, selection towards fitness, evolution into different species and so forth — that is, poliovirus obeys the same rules that apply to living entities. One could even argue that poliovirus undergoes sexual reproduction in the infected cell, as it readily recombines with sibling progeny or with other related viruses [. . .] to exchange genetic information[.] (Wimmer S6)

Wimmer is in fact attempting to limit the radical effects of his interpretation by splitting the virus in two, into a chemical (dead, computerizable, synthesizable) phase and a biological (living, evolving, and sexual) phase. If these distinctions prove deconstructible, if they lack the stability Wimmer presupposes, then he will be without a clear delineation of chemical from biological, dead from living, the sexual from the non-sexual, and of the informational from the physical. This was his initial claim — that a virus could be stored in a database and reproduced from its representation in bits *because it was a chemical*. Which means, in Wimmer's terms, non-biological, not life.

In other words, it is as if Wimmer is attempting to circumscribe certain of the more disruptive results of his interpretations by making them apply only to the "chemical," and by treating the virus as unique among "biological," evolving entities for going through a chemical phase. The ultimate effect of this chain of reasoning is to preserve a domain of pure life (the biological) that would be unaffected by the trace-effects that are at work disturbing the essence of virality and materiality. These divisions are tenuous. Any attempt to define the biological as the evolutionary will have to reckon with the applicability of such laws of heredity to chemical structures, most notably DNA (as well as cultural artifacts, ideas, and anything else). Whether all these evolvables are thereby living is a matter of definition whose arbitrariness can never be overcome. On the other hand, the characteristics Wimmer takes as definitively chemical, death or inertia, can always be found *within* the living. One could point to cellular and even multi-cellular organisms that can be revived after remaining frozen (and thus itemizable) for millennia, or recent advances in synthetic biology for evidence of this.[4]

More fundamentally, one could say that *all life is an economy of death* (Derrida, *Life Death* 272). By gathering calculability, materiality, inanimation, and informational translatability under the category of the "chemical," Wimmer attempts to circumscribe all the effects of the trace that have always been at work in and as life, making it possible and impossible. Life is not a special substance or power that takes its leave of a dead chemical precursor, nor is it simply reducible to a chemical understood as a logocentric formula — it has only ever been and been known, to itself or the other, to itself as an other, as a difference or differences that must transgress themselves to be re-inscribed or re-affirmed. There is nothing that can be called life that is not in principle calculable, yet the law or program of life is never known; there is nothing simply immaterial, yet only the future will tell what its material possibilities have been; life is not merely representable by chance within a variety of media and regimes of knowledge, including information technology, and then perhaps "resurrectable" from this representation, but from the first can only be instituted as a haunting trace that out-lives or sur-vives any given form. One could call this trace of life *virality*, the holding in suspense of borders of life-death and self-other.

That virality and vitality are only trace-effects, that there are only traces of virality and vitality, is one understanding of the displacement that Wimmer's experiment makes palpable — from the virus as a reproducing entity dependent on its "natural template," to a binary code stored on a hard drive. Neither of these forms is more true or essential than the other, neither for the virus nor the living, but the only "essence" is the interminable displacement that allows each of these representations and the doctrines of scientific knowledge from which they arise to become stabilized by the "same" forces that ultimately destabilize and supplant them. The virus is not reproduction or being-reproduced, being-copied, any more than it is information — rather, *virality* is that strange reproducibility and in-formability that allows for what-is to be repeated by what it most is not, which I am here calling trace and iterability. If the virus has always named something that troubles boundaries between life and death, because it only reproduces its self in the other and by perverting an apparently autonomous self-reproduction, one can see that it will have only been a name under which a futile effort was made to control the most disruptive, transgressive, or infectious effects of the trace by quarantining them under a name distinct from life.

Nor should the trace be confused with something like a vital force of creativity or invention that would allow life to exceed calculability and law of its own power, because of some innate self-overcoming. If the law of life is constitutively unknown, one will never be able to say when it has transgressed itself or when it is simply a different law that has been operative, which is already enough to cast this vitalism in doubt. By speaking of trace, I am seeking to avoid hypostatizing or substantializing this displacement or displaceability as a present and knowable subject from which the force of rupture issues and to which its effects return. The trace is no more subject than object, no more active than passive, no more law than the governed — even if there is something like a law of the trace, something obsessing or haunting us here by making it seem as if things-always-must-be-this-way, the trace is just as much singularity "itself," that which will never be simply brought under a positive law and which names the destabilizing of every legal authority, because it is this very inscription of the law, its derivation from what it cannot control nor account for, that makes every law possible while undermining it.

Thus, it is not because the chemical or viral is especially reckonable and programmable (as Wimmer suggests) that it winds up resembling a computer. On the one hand, there will always be a certain programmability to life, evolutionary laws can be run as algorithms just as easily as chemical reactions — this possibility of programmability follows from the same necessity that renders its law fragile and in a sense impossible. If this law is supposed to account for the entirety of history, then the most simplified as well as the most baroque and complex attempt to describe the totality of that history may be disproven tomorrow — but only by what in hindsight will appear as another program or programmability. And, if there is a programmability to vitality, there will be at the same time, following from the "same" necessity — the necessity that the trace is never simply the same — a certain inventiveness or rupture, a deconstructibility of the viral and the chemical. This is true in principle of anything "chemical," which is just as dependent on the *reproducibility* of the scientific method, the process by which hypothesis and experiment are centrifuged out of an iterability that leaves them as trace-effects, haunted by an alterity that never simply arrives.

This constitutive non-closure of the iterable trace is particularly marked in the case of Wimmer's chosen chemical, the poliovirus. Immediately after copying out its "empirical formula" as a quantity of atoms, Wimmer must concede that "because poliovirus is a *quasi-species*, the number of atoms in viral particles represents an average from a large population of different viruses. There might be little practical use in describing poliovirus by its empirical formula, but it persuasively portrays the virus as a chemical" (S3; emphasis added). The science of chemistry, the form it gives to discourse and to what we call nature, is here invoked not in the name of some truth value, but for the sake of *persuasion*, as a form of rhetoric. Certainly, the nature of poliovirus as a "quasi-species," as Wimmer and others understand it, follows from particular qualities of RNA viruses that are not universal among viruses, let alone chemicals. DNA can be proofread as it is replicated, so DNA viruses accumulate errors or innovations far more slowly than most RNA viruses. Quantitative difference becomes qualitative — rather than a genetically uniform population with a few mutants, it will be difficult to identify the prototype of an RNA virus infection, which will form an entire population of mutants whose original or essential form is more of an ideal, an "average," as Wimmer puts it. It is worth noting that, in a technical sense, Wimmer is misusing the term *quasi-species*, though his usage is perhaps dominant among virologists themselves (Holmes 91).[5] Whether in its exact technical sense or the looser sense in which Wimmer uses it here, quasi-specificity is certainly not a universal condition. Nonetheless, there is something like evolvability or deconstructibility, a quasi-transcendentality, that haunts every representation of biological or chemical identity.

One way of understanding this iterability, which allows the virus and everything else to repeat without remaining the same, would be to ask what makes the poliovirus the poliovirus. Obviously, it is not some genetic code, or chemical formula, because that is precisely what can and does change, at a startling rate. It may have other recognizable qualities, for instance its disease pathology, but all of these are just as dependent on their iterability and thus exposed to alterity. It is as if the only thing that remains is the one thing we know to be inessential, *the name poliovirus*, which is nonetheless open to the vicissitudes of linguistic evolution and translation and may transform without even changing a letter.[6] This is another way of saying that the virus must be *read.* There are no positive traits from which its identity could be derived programmatically or deterministically, not without the active intervention of a reading that can bring together or cast asunder predicates arbitrarily, without principle. This hollowing out of the body or content, the formlessness that is form — these are trace-effects.

Polis Virus

As they have been posed thus far, these trace-questions, questions that follow from or chase after something like a trace, might seem to be a relatively harmless or unserious intellectual play. Questions that take an ontological or philosophical form are often dismissed as inefficacious by scientists and even certain philosophers of science. No one ever succeeds in answering a question such as "what is life?" — those who attempt it succeed only in demonstrating the arbitrariness of their decision, admitting of rational critique but not rational justification. Moreover, it seems to make no difference whether we say that something like a virus or anything else is alive or dead, natural or informational, so why bother asking? Despite such appearances, these questions are always matters of life and death. It is precisely because no authority suffices to provide an ultimate answer that they must be posed each time anew. If we who call ourselves living do not know what the life in us is — what boundaries to draw around or within ourselves, and what causes this irrepressible emerging or urging — then no matter what we do, we will have acted *as if* we posed and answered this question of and to ourselves, however clumsily and provisionally. Nor do we even know if we possess the terms to ask "the" question, or if the question-form suffices, given that it can only borrow from the inheritance it presumes to place in question. Every act, even the most taken for granted or assumed to be impossible, implies an endless questioning or deconstruction of life and the good life, a displacement that is no more and no less ontological than ethical, political, aesthetic, and so on.

It requires very little speculative imagination on our part to find these domains already overlapping and infecting each other in the reactions to Wimmer's experiment. It is noteworthy that, although he did not publish his own theories and opinions about the ethico-political aspect of his research until several years after the experiment and the anxious response it provoked, he makes clear that he already hoped to address those questions in the initial publication but was prevented by the editors of *Science* (Wimmer S5). It is not even possible to say simply that the practical is already at work affecting the theoretical or scientific, but that, because this division too has only ever been the abeyance of a trace, one does not know where practical or theoretical begin or end, what affects what across what divide as their difference takes shape. Wimmer's experiment, even if it cannot simply accomplish something "new" (a claim Wimmer himself makes, compellingly), nonetheless brings about the effect of a displacement of borders. For instance, the knowledge and birth or production of the virus is brought under our control and thus our responsibility in a fashion that is at least previously unthought if not unthinkable. However idle may seem a question such as whether this computerized virus is the real or essential nature of virality, it is transformative for urgent questions in the fields that reckon with the viral — perhaps not the answer to that question (for one could say no and still be just as concerned about the possibilities of bioterror it opens), but *that the question be felt*, that its impetus shape the aesthetic or affective environment in which both "ontological" and "ethico-political" questions are posed. It is no longer a thought experiment or science fiction to ask how laboratory practices, scientific publishing and data-sharing, campaigns for vaccination and disease eradication, or military strategy and national security ought to change to acknowledge the viral possibilities implicit in this experiment.

Wimmer is eager to answer his detractors, who called his experiment irresponsible, but does not shy away from its disturbing implications: "Does the test-tube synthesis negate efforts to eradicate poliovirus? The conceptual answer to this is yes. Poliovirus cannot be declared extinct because the sequence of its genome is known and modern biotechnology allows it to be resurrected at any time in vitro. This is true for all viruses, including smallpox" (S7). Based on the definitions of extinction most common in public health, the virus can no longer count as extinct while it remains as a trace in memory or a memory supplement such as a computer hard drive.[7] It is always already a virus of the mind (a meme), a computer virus, and so on; it infects transcendence, corrupts its concept, and this non-self-identity cannot but transform the most practical considerations faced by a virologist, epidemiologist, or politician. Traditionally, the extinction of a virus would be considered a state of no return, and thus the permanent eradication of the associated disease (unless perhaps in the fullness of time the same pathogen evolved again from a near relative). However, even if the vaccination campaign underway to eradicate poliovirus is successful, meaning that there are no more infected individuals and thus no more transmissions of the disease, and on top of that if laboratory stocks of the virus were eliminated, the finality of this state is no longer secure. If it is possible to manufacture the virus from a genetic sequence stored in a database, there will always be a risk to ceasing a vaccination campaign. Its threat as a weapon only increases as a population loses immunity. The relatively practical, ethical, and political questions of public health and epidemiology are transformed in the wake of the changing onto-technological possibilities of virality.

A transformation of responsibility spreads like a contagion to everyone whose decision-making touches on or is touched by this new or newly understood possibility. A public outcry followed the initial publication of the poliovirus experiment in 2002; Wimmer attributed this reaction in large part to the heightened anxiety over terrorism in the wake not only of September 11th but also of the anthrax bioterrorist incidents that same year (S5). Wimmer drew a lesson from this response for scientific researchers and publishers, who he felt could inoculate against such anxiety by treating this ethico-politico-historical context as part of their scientific work and publishing their results along with analyses of these implications.

The company that sent Wimmer's lab the mail-order DNA segments necessary to compose poliovirus said, in the wake of the controversy, that they would begin screening ordered sequences against a list of known pathogens (Pollack). Such screening is now a customary, though by no means uncontroversial or flawless, practice (DiEuliis et al.; Gómez-Tatay and Hernández-Andreu; Piper).

Nor were the effects of transformed responsibility in the field re-oriented by this event limited to those strictly within the domain of scientific practice. Political decision-makers were involved on both sides of this debate. A Republican congressman representing Florida gave a characteristically duplicitous condemnation of the political and military implications of this research — in a draft resolution that never made it to a vote, he condemned the scientists conducting this research and the publisher but never mentioned the primary funding body, which happened to be DARPA (the Defense Advanced Research Project Agency of the Department of Defense) (Wimmer and Paul 592). This role of US national security in supporting synthetic virology exemplifies the undecidability of what is often termed *dual-use* research. Without pretending for a moment that the US government is anything but a global aggressor,[8] even if we imagine the most dove-like society conducting such research purely for purposes of defense (whose defense? Against what? The order being "defended" or preserved may already be the greatest violence), the *same* experiments that help us to understand and anticipate security threats provide knowledge and technology that can be used for ill. If some nation, no doubt imagining itself to be on the side of the good, were to attempt to conduct such research in secret, keeping classified all the knowledge it produced, the relatively private or securitized domain in which that knowledge were stored would necessarily be vulnerable to something like a *viral* intrusion — the reading and even re-writing of its code by some other, in a domain in which no priority or identity can be maintained between the "computer" or "biological" virus.

In truth, all knowledge and know-how are "dual-use" or undecidable — *pharmaka*. Everything that provides some power or potential for predicting and controlling matter or life necessarily grants a force that can be turned toward the best or the worst and will even risk destabilizing any certainty as to which is which. Wimmer uses an apt term when describing this state of affairs, one that captures the role the trace plays in deconstructing ethical as well as ontological judgment — research undertaken with the best of intentions nonetheless offers a "blueprint" for bad actors (S8). It is not simply that something originally good can be copied for the bad, but that nothing is in itself either good or bad — it can always be appropriated for better and worse. Nor is there "pure" research that would escape this problematic, but only research that is of unforeseen applicability. Wimmer offers the somewhat optimistic rejoinder that the free, public circulation of information and the self-regulation of scientific bodies is the best protection against the deviation or deviousness of the trace (S8). If I sound a skeptical tone in response to this optimism, it is not because I believe secrecy or "external" regulation is a better solution. Rather, it is only to maintain a vigilance that is necessary because this scientific institution is just as corruptible in its aims and its foundation as any other. Scientific authority cannot guarantee its own beneficence for the same reason that no external oversight is necessarily better — every authority can authorize the best or the worst and has no way of anticipating how its own work will be appropriated, even by its legatees. All of this is true even before one considers that, in the context of the United States, just to take an example close to hand, there is no way of framing the inside of the scientific institution that does not acknowledge the outsized role played there by some of the worst-faith actors imaginable — for instance, the US government and military, academic institutions, and pharmaceutical companies. These institutions have routinely proven that even the most innocent- or beneficial-seeming research can be co-opted for profit and the interests of empire. There is no simple opposition between biosecurity and bioterrorism — certain countries' control over medical technology and military force is precisely a reign of terror for the rest of the world.[9] It is not simply, then, the possibility of bad actors appropriating research done for the best, but a more fundamental instability of the notions of good and bad, which haunts the inside of any scientific research or decision-making, and which no virologist can insulate themselves from, even with latex gloves or a hazmat suit.

A similar debate surrounding *gain of function* research has intensified in the wake of the COVID-19 pandemic (Brouillette). Even independently of debates over the origin of the virus, the pandemic provides a daily and massive reminder of the potential risk of any research that increases the human-specific virulence of a pathogen. It is necessary for every stakeholder in this research — which is to say all of us — to weigh those risks against possible benefits, in a fashion that overwrites the ostensible boundaries of the scientific institution. In the case of gain of function research, which can include the cultivation of what are called *possible pandemic pathogens*, the justifications put forward seem like particularly ad hoc rationalizations of science being done without foresight or simply because one can.[10] It is said that this research will improve the surveillance of emerging viruses, because it can give us advance warning of which mutations will allow viruses circulating in wild or domesticated animal populations to jump to humans. However, the regular testing of viral genomes that would be necessary for this surveillance to be effective simply does not happen on the scale that would justify this reasoning (Lakoff 118–39). It is also argued that the process of preparing vaccines that might even stop a pandemic in its tracks can be expedited by this research. However, here we see more clearly how the social and institutional structures surrounding scientific work overdetermine its significance and efficacy; many have pointed out that even if vaccine preparation is expedited, our systems of distribution prevent the global distribution necessary to effectively stop a global pandemic, particularly because distribution is determined by for-profit companies (Evans et al.). The COVID-19 pandemic provides a disturbing reminder of the injustice of this system, as much of the world is still without access to any vaccine (Dearden). Here, we see how the theoretical and practical significance of scientific research and the responsibilities "internal" to the scientific institution and its practitioners are shaped by the social, economic, and political framework in which it takes place.

Even if, within a certain context, such decisions appear straightforward, in principle they open on the unanticipatability of the future, in a fashion that calls us in each instance to make a decision that cannot ultimately be justified. There can be no principled response, no responsibility in principle, in the face of a threat that has no simple presence, a threat that is the opening of a future as such. That is certainly not to say that everything should be permitted, but only that every such decision requires an each-time-singular responsibility, which could not be derived from a principle or program or even rationally justified. It will always be possible to give justifications, and one is even called to do so, but there are necessary or structural reasons why no such justification or rationalization can suffice. It follows from the *same principle* that any research, or any prohibition of research, may have unforeseen benefits or do unforeseen harm. Thus, no individual or body entrusted with evaluating research can provide an ultimate or authoritative decision on these possibilities — the very process of science is open to a sort of insight and even a structural unknown or ignorance that no one controls. Thus, even before one confronts the necessary possibility that any one of these individuals or agencies is corruptible, that financial, careerist, and political influences may lead them to obscure what they know or intentionally advocate the worst — even relying upon the best possible faith, there is no certain measure of what may come. There is no uncorruptible or infallible agency — whether such research is evaluated "internally" or "externally," by the scientist conducting it, a scientific body, a journalist, a governmental organization, nationally, or internationally — which does not at all discredit such oversight, but rather intensifies the vigilance that no one can absolve themselves of precisely because none of us alone or collectively can meet its demand.

The complexity of this scene of responsibility, which touches or infects us all, has been acutely felt throughout the pandemic. Each of us, whatever power we have to act on the outcome, can feel that everything from our most minute daily activities to national policymaking is connected to a science-making that is still in process and whose results are anything but given. It is never as simple a matter as saying that one should "follow the science" (even if this may be an apt political slogan in the face of willful disregard), because there will always be disagreement within the community of those calling themselves scientists, as well as the possibility for all kinds of bad actors. One indulges in the pretense of endowing another with a decision made by the self when one pretends to "trust" science, while in fact *reading* it from a scene in which one is necessarily left to decide which science, which scientist or science communicator, which journalist or politician merits this faith. Anyone communicating science for the public or simply in public must make decisions, not just about who is a trustworthy representative of "the science," but how issues should be framed, which consensus and which disagreements get aired, what merits being treated as established and what as subject to reasonable debate, and what ought to be left unmentioned. There is no line neatly dividing theoretical and objective knowledge from questions of values or action, nor is there a clean distinction between the science itself and contextual factors determining how to present it, given that the decision of where to draw this line is precisely what is in question.

My objective is not simply to itemize the widespread displacement of borders that occurred in response to these viral possibilities, but to show that everything taking place here is a trace-effect. The virus is no more ontological than ethical, no more scientific than political, because these domains as well are formed as the remanence of so many traces, as evidenced by how quickly Wimmer's relatively circumscribed experiment could transform the projects and responsibilities of global public health, scientific research and publishing, and politics or national security. This re-orientation does not take place because a virus, as some present and known entity, crosses or disturbs existing borders, but because a *virality without limit* first comes to be as the very displaceability of any and every border.

Proof of Principle

There were also detractors of Wimmer's experiment who criticized him not for doing too much, but too little. They said, not that he had opened the door to bioterrorism, but that he had executed a mere publicity stunt. To respond to these two classes of critics, Wimmer pursues somewhat conflicting strategies. I am less interested in catching him in a contradiction than in attempting to recognize a fundamental aporia of scientific "discovery." On the one hand, in response to those critics anxious of bioterrorism, Wimmer insists that his experiment accomplished very little, almost nothing. It took something that was already possible (necessarily, or it wouldn't have worked), and simply actualized it. He created no technologies nor even technological applications that were previously impossible, and any terrorist or ill-intentioned actor could have done what he did before him or without him: "All methods used for the synthesis of poliovirus were published long before the experiment was conceived. Thus, we neither described new technologies to synthesize DNA nor invented novel methods to convert cDNA into infectious viral RNA. The purposes of the poliovirus synthesis — namely, to establish a proof of principle and to sound a wake-up call — were accepted as reasonable at the 2003 workshop [on 'Scientific Openness and National Security']" (Wimmer S8). Already its "first" time, nothing more than a "proof of principle." Here too, a certain virality precedes us; André Lwoff recalled that some early virologists referred to the virus merely as "the principle" ("Lysogeny" 273). There is never anything simply and straightforwardly new. Even for those researchers who first achieved artificial DNA synthesis, or any other technology that allowed for a transformation of scientific practice, the "novelty" of the results made possible by its permutational elaboration can only appear on the basis of the now old-fashioned or outmoded regime of research. The same is true of those relatively "theoretical" inventions, "new" interpretations of existing research that chart an unanticipated course through their field and may allow for a certain effect of pathbreaking or *frayage* in the process. This state of affairs, which I doubt would be denied by any historian of science, but which may nonetheless be dismissed by some as a truism without effect on this practice and its efficacy, nonetheless leaves a haunting dis-jointure in every apparently positive statement of the state of a field of knowledge. We will see an example of this when we come to the "discovery" of the virus, which only appeared as a gap in the field of knowledge opened by germ theory and the Pasteur-Chamberland filter, and never has settled nor will settle into a state that is not a displacement of a stopgap structure.

It is this necessary incompletion of scientific practice and knowledge that leads Wimmer to state a precaution: "It must be emphasized, however, that we have merely reproduced poliovirus by following the blueprint of the viral genome. We did not 'create' this virus" (S6). He draws this distinction between *reproduction* and creation or *production* because he acknowledges that it would not be possible, on the basis of current knowledge, to perfectly anticipate the efficacy of a novel virus. Too little is known about what allows a given structure to succeed in infecting a given host, and thus this "mere" reproduction is all that even a synthetic virologist is capable of. That every known virus is only the derivative of an unknown principle is another way of expressing the dependence of knowledge and being on something like a trace. We do not yet know what a virus is or can be — or whether it is, whether that thing we call virus is *one* — that will always be true, no matter how manipulable and manufacturable they become. No one will ever create a virus without a blueprint or model (otherwise, by what right would it be called *virus*?), and that model will never be anything but contingent and deconstructible, dependent on a provisional apparatus of knowledge and technology. That there is no *pure production* is the virality of everything. There is only birth or invention that depends on some other, on what it is not and cannot recuperate or re-appropriate having been there already.

Despite or because of this trace, which is neither simply a limit nor an empowerment, neither repressive nor productive, even the most apparently derivative repetition may bring about effects of invention and discovery. What we call invention or discovery can come about in no other way. Wimmer's experiment is no exception to this rule, and it is these effects he seizes on when he hopes to counter those critics who accused him of staging a publicity stunt. Despite saying, earlier in the same article, that he had done nothing that could fall into the wrong hands, because he had in fact done almost nothing, Wimmer nonetheless points to positive effects of discovery and invention that followed directly from this experiment: "Although some critics claimed that we did not learn anything, the poliovirus synthesis in fact yielded valuable scientific information about the genetics of viral pathogenesis. Indeed, our experience of synthesizing poliovirus has inspired us to pursue viral genome synthesis for medical applications, for example, to explore novel strategies in vaccine development" (S7). Wimmer included a signature in his synthetic virus that would allow it and its progeny to be distinguished from "wild-type" virus (S7). He substituted certain codons in its genetic sequence with synonymous triplets. That is, the translation of RNA into protein operates by means of a code or alphabet that converts triplets of nucleotides into one of (typically) twenty amino acids. There are more possible RNA triplets than there are amino acids to code for, and thus many amino acids have more than one triplet that codes for them. Wimmer's substitutions made no difference in the protein structure of his resulting virus, yet his team found that their synthetic poliovirus had significantly reduced virulence. They attributed this difference to these substitutions, though the mechanism of such attenuation is still under study (Wimmer and Paul 598–603; Groenke et al.). Wimmer now holds several patents on viruses attenuated by codon pair bias, which has become an active field of research for vaccine development (Coleman et al.; Tan et al.). Every "invention" takes form in this way, even if certain of the empirical details have a different cast.[11] The unanticipated and unhypothesized happens despite everything occurring within the domain of positive knowledge as a mere repetition or reproduction. From one perspective reproduction or reproducibility fails — from another, a principle that has always operated has yet again re-instantiated itself, unbeknownst to the experimenter. One moves from one unknown to another, or faces different aspects of the same unknown, within a domain where everything resembling control and creativity (reproducibility and invention) is derivative of something that has no power and does not exist. One only invents viruses, which means one never invents. One never knows what has been invented, or what has invented — behind the scientist some virality is operative.

Viral Matter

A virus is no more biological than chemical, no more scientific than political or economic, no more natural than technological, synthetic, or artificial, no more discovery than invention. It is none of these things because there is nothing it simply *is*; it does not belong to being or the ontological any more than to ethics or epistemology. I focus this study on the virus because of the disruptions it reveals in the theory and practices of life and the living. Not because the virus, as a particular well-defined essence or entity creates and destroys borders, but because it has only ever been a way of speaking about a *virality* without positive term or germ that comes to pass as the stabilization and destabilization of the circumscriptions of the living. The virus deconstructs — which is not to say that it is an agent or power but that it exists only as deconstruction.[12]

Thus, there is no *materialism* that could contain the operations of the virus, any more than a vitalism or any other ontology. The problem with such an approach is not exactly the *material-*, which may well be one face or phase of the virus, a mask it wears, behind which it practices its subterfuge, but the *-ism*, which means, if it means anything, that thought follows a doctrinal course, trying to ground everything from a first principle or cause. Virality disturbs any effort to bring nature or being within first principles, not because it arrives from outside the origin (in which case it would remain defined by it, respecting its limits), but because it simulates originality itself without ever simply belonging to it. If it is possible for a virus to pervert the course of the origin, that is only because the origin is made possible by an iterability that leaves it a ghost of itself, opening it to diversion from before its "proper" beginning. The only *ism* appropriate or adequate to the virus would be a *seism*, a trembling of foundations and the boundaries erected on them (Derrida, "Some Statements").

For precisely this reason, because the "origin" governs nothing and is everywhere reappropriated by some alterity, it will be possible to speak in the name of materialism or ontology and say anything at all. It would be naïve to approach the field that designates itself today on the basis of a "material turn" and expect to find a fundamental continuity of principle, method, objects investigated, or anything else. Nonetheless, certain trends or motifs can be found at work in the field, even if they only represent one strain among others, and often only a single stratum within a heterogeneous and self-contradictory text.[13] One such tendency turns to the field of inquiry of the natural sciences and accepts what one finds there as the power and agency of "matter," apparently without mediation. This tendency can be identified even in those texts that argue for a fundamental "entanglement" of "natureculture," while nonetheless insisting that this hybridity *belongs to* matter or nature itself.[14] The problem is not so much that the "material" or "natural" is chosen as a name for the grounding substance or subject, but inheres in the very idea of *belonging*, which treats appropriation as the act of a sovereign subject to whom all inventiveness or transformative power returns. Every invention may be the invention of the other. Ironically, it is impossible to claim that the pure agency of matter is an object of our knowledge unless one effaces the contingency of our representations, treating language, thought, or science as a kind of absolute. If we entertain the possibility or perhaps the necessity that what we call "matter" remains in some way unknown to us, then no "materialism" can forego tarrying with and placing in question the language and history of its categories. To restructure this field starting from such a complicity is not to posit a language or culture that would be prior to matter or nature, but to recognize that no priority can be discovered or named within this circle in which we find ourselves turning.

Deconstruction is not a particular methodology, not a way of thinking or writing about an academic field. If a text may *deconstruct*, and if it may even be in a certain sense necessary that this come to pass, that is only because deconstruction is *what happens*, is the only possible chance or opening for an event that is not at all opposed to or kept pure from its contamination with something like textuality (Derrida, "Time" 17). If we would like to leave space for the most irruptive arrival of the unthought, far from simply turning to an established matter or nature from which the new would follow apparently just as it always has, it is necessary from the first to confront the fact that *matter* and *nature* and anything else we could put forward to feign knowledge of the origin are necessarily *words in a given language*. Acknowledging this obvious truism, which nonetheless is everywhere suppressed, is as far as possible from committing oneself to a linguisticism or idealism that would pretend there is nothing outside language or thought in the traditional sense. Rather, it is only possible for that beyond or that otherwise to shine through if the presupposed limits of language and thought are destabilized. Rather than simply being the sovereign inventiveness of a given subject or power, there is no event that would not come to pass as a re-inscription of iterable and thus displaceable boundaries — that is, as deconstruction.

Deconstruction would not be reducible to another approach to the humanistic study of science, which attempts to ground scientific thought in the language, culture, history, or subject position of the practicing scientist. This approach, sometimes identified with the "strong programme" (Bloor), and prevalent in the sociology of scientific knowledge and science and technology studies, may be the apparent opposite of "materialism," though it is just as possible to find its methods and presuppositions at work in the field that describes itself as materialist today.[15] One can try to rationalize this by speaking of a matter that forms culture in order to re-form itself, but it is just as possible to recall that once a term like *matter* or *materialist* exists it can be made to refer to anything at all, without requiring that its referents share anything like a common essence or origin. There may be no essence at all apart from this simulation of it by an open-ended iterability. "Materialism" is no more a re-grounding of this field than a simple denegation of its inheritance — there is only such unfaithful, "inventive" heredity, which is why nothing ever simply returns to the power or agency of a given subject (material or otherwise). Rather than treating matter or life as self-evident facts of experience, this sociological approach begins from the recognition that life, for example, *is a concept*. This concept has a history, it is anything but natural, and one turns certain scientific discourses on their head when one investigates how this concept becomes tied to a certain regime of experimentation and interpretation. Evelyn Fox Keller carries this line of inquiry to its natural conclusion, that "the category of life is a human rather than a natural kind" (*Making Sense* 294). This would mean that *life* is no more than a word, a cultural product forcibly imposed on an unwitting nature.

Whether one identifies the "materialist" field with or opposes it to this tendency, which is sometimes gathered under the heading of *constructivism*, I isolate it here to try to open the space for a deconstruction that would happen otherwise. Neither as material nor linguistico-cultural power. Deconstruction is not what happens when a language or culture *imposes itself* on matter or nature.[16] Such a story preserves the presuppositions of subjectivity and the fantasy of sovereignty. It narrates as if a language or culture had been there already, and remained what it was regardless of the differences it incorporated or overcame, acting with pure autonomy upon something separate from it (the material-natural substrate whose original self-identity is also feigned). Rather, deconstruction is the very movement in which the supposed purity of origins is already caught up and carried away, happening in its "first" occurrence as an inscription or infection, a virality whose contingency is its only necessity and that leaves the originalist with nowhere to *turn*.

Deconstruction displaces these schemas not by choosing a different ground but by placing in question the value of origin or production that insists there be a cause, whatever its name. The first time is a repetition. As a result, we *must* make an intervention into a field already polarized by these oppositions *without* ever having access to a true ground for their partition. Being left to decide, where no reason will suffice and yet where no other could decide or give reasons for us (at least not with ultimate authority) — passing through this impasse is what Derrida called deconstruction.[17] We must make life-or-death decisions without knowing what life is. What I am proposing here is equally a deconstruction of the life sciences — and the life sciences *as* deconstruction.

Deconstruction is more than a disorder of certain attempts at theoretical position-taking. It is just as much at work where materialist or realist strategies frame themselves as responses to an event or impending catastrophe that requires ethico-political action. This is often the justification given for theorization or ontology undertaken as part of a "material turn" — that it will provide a necessary corrective to the habits and patterns of thought that have led us to ecological catastrophe, and will help to imagine strategies for a better future. This catastrophe includes not only the climate crisis, but also the ecological conditions that are incubating future pandemics (Malm). Far from dismissing the urgency of action and the demand for responsibility within this field, deconstruction is the only possible intervention in a field where truth proves insufficient. Without for a moment casting doubt on the urgent need for a transformation of our political life and its relationship to the environment or environmentalism, it is possible to question certain of the framing gestures that have attempted to define the impetus for this intervention. Despite the progressive political agenda of most authors in the materialist field, there is the risk of a certain conservatism in how this framing seeks a unified cause and origin and thus a single discourse or authority accountable for an adequate response. By turning to "matter" or the "nonhuman" to find what has been left out of our demos, and by attempting to define the present "age," whether as an Anthropocene or under some other heading, to represent the earth in toto brought under the sway of a unified cause, one simulates the self-identity of causality and the agent capable of redirecting it. This "materialism" is often accompanied by an investment of authority in the discourses of the natural sciences — it is not without relation to a common pattern that leads liberals to embrace conservative ideals in times of crisis. This pattern has been particularly observable in the US context since the 2016 election, though it was already in motion long before; patriotism, nationalism, militarism, family values, and other conservative commitments become rallying cries for many liberals who are more invested in restoring a status quo than challenging its foundations. In a similar fashion, it may be comforting in times of great risk or uncertainty to imagine that an authority exists with unique access to the truth and a plan of action for an adequate response, in much the way that "follow the science" has become a rallying cry throughout the same period.[18]

By questioning this attribution of authority, I am not suggesting that there is some other cause and some other authority responsible for ecological stewardship — and certainly not that any cause can be captured or action taken without some fictionalization of its unity. Rather, there is only the deconstruction of responsibilities and the divisibility of origins. This divisibility is already well underway in -cene discourse, which exposed the differentiality of origins as soon as the new geological epoch was put up for consideration.[19] -Cene itself has become a technology, an easily iterable and portable morpheme that elevates everything it comes in contact with to the level of a or the fundamental historical force. All of these terms are capitalized, treated as the *proper name* of an unsubstitutable reality. Nonetheless, the effect of multiplying these epochs or nomenclatures is ultimately to divide the figure that is placed at the origin as cause, and thus to suggest a different course of action to right the trajectory of history — the Capitalocene, the Plantationocene, and other neologisms have been used to suggest that there is no unified "Anthropos" that is collectively responsible for the environmental state of the planet, but that specific economic, political, and racial systems of oppression are root causes.[20] It is not humanity *as a whole* that has destabilized an ecological balance, but the class in control of a capitalist system that demands constant growth and environmental extraction, or the racial and national formations that have globalized this system through colonialist and imperialist violence.[21] Such displacements do not simply suggest a different historiography, but challenge the unifying representation of a collective humanity who bears responsibility for the past and can control the future. They also suggest, by turns, the academic discipline or disciplines best suited to guide this movement. The problem is not that any of these analyses are wrong but that *each in its own way is correct*. I would challenge not the analysis of the present moment on the basis of categories of species, class, or race, but the -cene, the implicit demand that these analyses lead to a single, teleological narrative of historical development. Any category we extract from that history and place at its source will necessarily seem both to have always been there and to have undergone displacements both subtle and massive throughout a history of which it cannot simply be the organic origin and unity. Deconstruction would not be a selection from among this field, of the true epoch or its true cause, but the very movement by which that cause places itself in question, dividing from itself without providing any simple resource for its narrativity.[22] Deconstruction is the displacement of every -cene, not in order to halt the transformations of responsibility in this field without closure, but to suggest that the agent bearing responsibility for the whole will be an *outcome* of this movement, rather than its origin.

We do not know where or when we are, or how to decipher the trajectory that determines the way forward. Every such debate resembles a *battle over the proper name*. In this respect, it is not only the -cene or Anthropocene, word or thing, that is in question but just as much apparently epochal terms such as "climate change" or "climate crisis," which risk effacing just as many differences when they are made to stand for *the* problem of *the* present epoch (it is worth remembering the all-too-political negotiations that produced the term *climate change*). I am not simply trying to chart the shifts of discursive or academic norms if I suggest that these terms too may come to seem outdated long before the movement they are attempting to name and the forces they are trying to harness have been tamed. It is not because something ceases to exist that a name falls out of use; to understand this drift better we could compare the discursive scene that is only a few decades past, in which *overpopulation* was a buzzword attracting a similar anxious energy and apocalyptic imaginary. It is for good reason that this term strikes most as outdated today — it shifted responsibility for overconsumption from wealthy nations to those parts of the Global South with higher birth rates, often with explicitly eugenicist politics.[23] Still, it is not that "overpopulation" does not exist whereas "climate change" does, but that heterogeneous and conflictual forces are differently framed by these implicit discourses or narratives — forces that have no phenomenal or ontological presence without this placement and displacement within and between narratives. The reality of the real is not what exists or persists outside of these textual effects but what a certain contextuality privileges over other framings. Certain continuities can be recognized between the framings of "overpopulation" and "climate change" — many things that we are experiencing in the present and expect to worsen: famine, drought, pandemic, and environmental degradation, were predicted by both models. I am not at all trying to vindicate the discourse of overpopulation, but to help us to understand why a similar process is happening and will happen to our present terminology, whether "climate change," "Anthropocene," or anything else. That is, it is not a matter of a simple or pure falsehood or nonbeing that makes a framework obsolete — its fictionality is our only resource, we can only oppose one fiction with another, and for that reason it will remain re-appropriable by the very forces it attempted to counter. If it is up to us to sort activism from greenwashing and the grassroots from the astroturfed, that is not because we have chosen the wrong framing but because any attempted circumscription will remain re-appropriable. The process that is already underway by which the same account that has made the crisis recognizable has nonetheless obscured it — has led to massive yet insufficient political and economic shifts that would not have happened otherwise, while also allowing what can from a certain perspective be recognized as the very system in question to rebrand its operations without changing them — this process does not take place because we have chosen the wrong name but because any name will be similarly co-optable and will require the constant vigilance of those who deploy it to mobilize against the worst.

One can even foresee how this might end. Conflicts are already emerging within this paradigm that may ultimately overtake it. It is entirely possible, long before greenhouse gas emissions are stopped and before carbon has been recaptured from the atmosphere, that the very ecological and political urgencies the paradigm predicted will undermine "climate crisis" as an effective and unifying *name* for the exigencies of the present. Certain signs of this transition are already legible in the present. As more and more of the world faces thirst, starvation, and displacement, as more tipping points are crossed, the political project suggested by climate politics may increasingly come to seem like a luxury reserved for those who have a future to count on. Concern for the green, the environmental, the climate, may be wielded against those who are forced to demand basic resources in the present and who challenge the politics of borders. A time may come when the very invocation of a "climate crisis" will be just as politically suspect on the left as "overpopulation" is today (and today's most vanguard discourses will sound just as stale) — even if successor rallying cries unite around a vision that has much in common with the politics once thought under the former name. This will not be because the problems it named no longer exist, nor because it was the wrong heading for political mobilization in the first place, but simply because a context has shifted around it. I raise the specter of this future only to remind us of the undecidable relations of word and thing even or especially in the face of the most urgent necessities.

Everywhere that diagnoses and cures are put forward from a "materialist" perspective, a certain faith is placed in the efficacy of the truth. Even when it is said that something like scientific communication does not suffice, that other modes of knowing or ways of feeling would overcome our current impasses, for instance that literary representations of nature in crisis will allow us to "imagine otherwise," a certain *critical* hope remains, that one will be able to identify the good *in theory*, as an ontological and ethical position that could guarantee resistance, beneficence, or innocence. These materialist and ecocritical studies often frame the worsening ecological disaster as an effect of a lack of understanding or erroneous mindset that fails to grasp the consequences of its own actions and our own ultimate dependence on ecological stability. Thus, these theoretical interventions are cast as a consciousness-raising that is expected to lead to political formations adequate to respond to the root causes of environmental degradation. While any number of examples could be drawn from this field, I will cite just one instance from Jane Bennett's *Vibrant Matter*, where it seems to me that the call to grant political representation to the "nonhuman" is circumscribed within the imaginary of the bourgeois-liberal subject: "We are, rather, an array of bodies, many different kinds of them in a nested set of microbiomes. If more people marked this fact more of the time, if we were more attentive to the indispensable foreignness that we are, would we continue to produce and consume in the same violently reckless ways?" (112–13).[24] This project of consciousness raising is ultimately guided toward a classical ideal of the self-possessed subject of representative democracy — despite everything in Bennett's work that speaks of networked interaction, her vision of political agency remains centered on individual conscience, consumption habits, and the vote or "voice": "But surely the scope of democratization can be broadened to acknowledge more nonhumans in more ways, in something like the ways in which we have come to hear the political voices of other humans formerly on the outs" (109). I have several reservations regarding this analogy between human and nonhuman (Bennett follows it with a citation from Latour that compares our lack of political representation for nonhuman things to the denial of the vote to "slaves [*sic*] and women"), but I will focus on two limits it imposes on her political project: (1) political practice is imagined within the sphere of the free speech of self-present, self-conscious subjects, where (in keeping with the conviction of the efficacy of the truth) the best ideas win out.[25] In other words, it is assumed that the means to realize a political project simply exist within the present status quo, and we only need to affirm our commitment to a goal to realize it. (2) Without entering into the immense problems of who will *speak for* the nonhuman or translate its voice or vote, we can recognize the limits this phonocentrism places around the scene or staging of political agency. Bennett's and Latour's analogies between Black rights, women's rights, and nonhuman rights relies upon a false historiography in which the voices of these *once* excluded groups *have been heard* in the present.[26] There are as many signs today, which I could not exhaustively analyze here, that the formal "inclusion" of these voices has been little more than a tool to feign collective consent for socially, economically, and politically inscribed limits that cannot be challenged by our electoral system.[27] Thus, the liberal politics that always emphasizes voice — bringing more people into conversation, dialogue, discourse — remains circumscribed by structures of power that overdetermine that discourse's efficacy.

Without for a moment disavowing the importance of education or consciousness-raising in any of the forms it may take, I would place in question the implicit model that guides this conception of its subjects and its political efficacy. According to this representation, learning about or feeling with matter leads to an appreciation of nature and to political formations capable of its appropriate stewardship. Whether this conversion is pictured as rational or affective, a very real possibility is not faced, one that I fear accounts for a significant amount of what we call *climate denialism* — the possibility of desiring destruction and self-destruction for their own sake.[28] This possibility is often suppressed because of the instabilities it introduces into political discourse and into the very conception and self-identity of the subject — who is capable not only of harming themselves but also of *desiring* and even *taking pleasure in their own pain*.[29] It may be that someone who enjoys consumption, who feels that it feels good, is simply unaware that they endanger their own life or legacy in the process, but it is just as possible to willingly do violence to oneself and one's dependents. The psychic byways that lead to such actions can be convoluted, but may amount, for example, to neutralizing a vulnerability by torturing oneself in order to beat anyone else to it. This psychic bargaining is capable of compromise and subterfuge to the point that it is not even necessary for a person to do violence to themselves (or their environment) to feel satisfaction, but merely to identify with their own exploiters.

Acknowledging such a possibility, which may be the dominant drive or ego defense of American politics and American life, is by no means a way of discrediting in principle projects of education and consciousness- or sympathy-raising. I am certainly not suggesting that it is better *not* to educate — rather, I would suggest that for much of the target audience of this project, the truth is already known. In order to feel this satisfaction in one's own destructive habits, it is necessary at some conscious or unconscious level of the psyche that the meaning of one's actions be understood. For those forms of conspicuous and useless consumption sometimes called *petro-masculinity*, which could be described as carbo-phallogocentrism, it makes no sense that they would be desired at all unless their destructiveness was intuited (Alaimo 91–110; Daggett).[30] The justifications that are sometimes given, including climate denialism or a desire to spite perceived hypocrisy ("owning the libs"), are just that — rationalizations, justifications invented after the fact by someone who may not be willing to face their own drives. Thus, it is not at all to discredit the projects of writers and educators that I place in question the efficacy of the truth, but to reconceptualize that project. Education is not the imparting of effective information, but the working through of a resistance. I doubt that I am saying anything that would come as particularly surprising to educators, who I would expect at least intuitively have recognized this as part of their work for as long as they have been in the classroom, or any other site where learning and transference take place, where such resistance is faced. They will also recognize that there is no way to program a successful working-through — the byways of psychic investment and disinvestment are circuitous enough that the scene of teaching will remain an unanticipatable event (Bojesen).

As I have tried to make clear, a deconstruction of this field is not at all *against* the truth. It is not a matter of placing in question what we call the scientific consensus on an issue such as climate change or replacing it with some sort of fiction or counter-truth (in the limited sense). Rather, it is a matter of recognizing that what we call the truth takes place in a context that is governed by forces more powerful than it, forces that give it its form while circumscribing its efficacy. From the myriad ways that one could observe economic and political forces shaping scientific work, its dissemination, and its re-incorporation into (or rejection by) economic and political comportment, I will focus only on the aspect that seems most relevant to the framing of the political project of new materialism. As is the case with every academic specialty, when one turns to the materialists one finds the most conflicting and heterogeneous tendencies at work, often within a single text. It is avowed that one must see oneself as dependent on and interwoven with all of life and matter, and at the same time that the future of this fabric depends on decisions that will be made *by us*, whoever we are — that our political *demos* must be broadened to include all of nature and that this can only happen by *finding it in ourselves* to recognize the agency of matter. I am not suggesting anything like a novel solution, as if we could step outside of the circle drawn by these internalizing and externalizing tendencies that are nonetheless not exactly opposed or opposable — deconstruction is not the exit from a system but takes place as a re- or dis-orientation of its forces. Even if systemic transformations that incorporate all of us are impossible without at least certain individuals undergoing something like a change of heart, it is necessary to be suspicious always and everywhere of any discourse that imagines a subject without self-contradiction guided by truth and sympathy. These self-realizations only take place within a context that ultimately circumscribes their efficacy. It is even necessarily possible that individual proclamations by those in power of their dedication to this soul-searching may serve as an alibi while leaving this system untouched, or that those without such power "work on themselves" in lieu of the systems they cannot alter. It is not possible simply to bring this context within our calculations, to reduce its effects by reaching the telos of a social whole with which the individual will can be reconciled, but one can to varying degrees use relative effects of context to transform the possibilities of the "self."

A limit is set to what can be accomplished within the orbit of one's own conscience by the political circumscription of individual agency. Together with the appropriability of any position-taking or discourse, and the vicissitudes of the drives, this circumscription overdetermines what any ontological project can accomplish. Imagining that ethico-political commitments and actions follow directly from individual consciousness-raising not only plays in to status quo politics but also cuts the most urgent and overwhelming crises to fit the interventions available to academics in the most limited understanding of their social role (as teachers and writers but not, for instance, as organizers). To stay with the example offered by Jane Bennett's work, because her vision of political transformation is focused on these individual attestations of the will (she concludes her work with "a litany, a kind of Nicene Creed," which reduces its project to the avowal of a set of *beliefs*), they could only be enacted by the limited political and consumer choices available within the existing social framework. Any such discourse, that expects a political project to follow from a *theory* or *ontology*, a discourse of the truth, will be inevitably surprised by the ease with which every positive result of its science can be reappropriated by the very forces it hopes to oppose. Thus, in the American context, which is my own, any subject restricted to this private attestation of political will, facing their conscience in the solitude of the voting box, is faced in every general election with a bipartisan consensus that privileges the profits of capitalist extraction over the needs of the living.

Given that "materialist" theory expects a political will to follow directly from an attentiveness to the natural world and natural sciences, one should evaluate the uses that "follow the science" has been put to as a political slogan. Here as well, the appropriability and circumscription of such position-taking is apparent; ultimately, the Democratic party wielding such slogans is one flank of a political system that excludes the possibility of certain political actions, whether or not they are recommended by scientists. This does not mean that we, as political actors, should *disbelieve* "science," but that we must in our own activism or advocacy take pains to distinguish our work from a context that can so easily absorb and subvert it — particularly, by maintaining a skepticism of any framing that posits a secure shelter of truth within an existing authority structure such as scientific institutions. The Biden administration, which has continually invoked scientistic slogans, has remained just as committed to existing political and capitalist power structures in the case of climate politics, where it has sought "compromises" that will increase the extraction of fossil fuels, as with respect to the politics of the pandemic. For all the talk of a break with the policies of the Trump administration, Biden's promises to "follow the science" have given way to an *indistinguishable* politics and economics of health — in essence, a vaccine-only public health response.[31] His administration took every opportunity to diminish the severity of the virus and extent of its spread, to oppose lifesaving structural interventions in public health — which should at a minimum include universal health care, paid sick leave, improving indoor air quality, mask mandates, and testing — and to shift the burden of responsibility for understanding the virus and taking appropriate precautions to individuals (including, ultimately, even placing vaccine access in the hands of the private market). My analysis of this bipartisan consensus on the politics of public health is indebted to Beatrice Adler-Bolton and Artie Vierkant, whose podcast *Death Panel* and related writings have offered a relentless critique of the Biden administration's "sociological production of the end of the pandemic," while articulating an alternative, anti-capitalist vision of public health that recognizes its inextricability from political and economic transformation.[32]

Within the first year of Biden's presidency, the phrase "pandemic of the unvaccinated" began to circulate among administration officials, the director of the Centers for Disease Control and Prevention (CDC), and sympathetic journalists, shifting blame onto individuals for the death, disability, and disruption caused by the lack of public health response. The administration continued to blame unvaccinated individuals for the pandemic even as data made increasingly clear that vaccinated individuals were still spreading the virus and frequently becoming disabled or dying because of it. The CDC director Rochelle Walensky even blamed individuals for their deaths, arguing that these were acceptable losses because most of those dying had "comorbidities" (in addition to its eugenicist logic, an insufficiently evidenced claim). Until April of 2022, the CDC hid its data on deaths among the vaccinated from the public. During the omicron wave, which officials called "mild" and baselessly claimed would lead to "hybrid immunity," the CDC reduced its isolation guidance for those who tested positive from ten to five days, despite knowing that the infected could remain contagious for much longer. Economic and political concerns motivated this change, in particular keeping health care and travel industries staffed. In February of 2022, in an effort to justify the end of mask mandates, the CDC changed the metrics it used to map localized severity, weighing the availability of hospital beds over rates of transmission. Despite the fact that masking is far more effective when universally adopted, as mask mandates expired the administration and public health officials increasingly told people to base their decision to mask on individual evaluations of risk, even as they were taking away sources of information about transmission. The administration allowed testing sites to close and shifted to at-home testing, leading to a significant drop in the number of cases reported. In September of 2022, at an auto show, Biden would then point to the absence of masking as proof that "the pandemic is over" (Vierkant and Adler-Bolton, "Year III").

While promising to believe or follow the science, this administration has (inevitably) chosen what it would accept as science and promoted the careers of those scientists or science communicators who would stay on message. The necessary response in the face of this inevitability is not to pretend that an apolitical science exists that can steer us back on the right ethico-political course, but that one must fight on an ineluctably political terrain even for what will count as "science."

Both in the case of ecological disaster and pandemic response, there is no way within existing political structures to enact or act on the political priorities that an attentiveness to the needs of the living ought to demand. Every attempt to re-invoke the importance of science or nature risks appearing, if it is not introduced in a way that displaces the limits of this political system, as a piety of the converted or a hypocritical self-righteousness. I am not reciting this well-known state of affairs to suggest some sort of despairing inaction, but to insist upon the necessity not simply of distributing information but of a collective *organizing* that first makes the subject of such a transformation possible.

The science of life and the knowledge and power it offers to the living do not simply provide, as a corollary, the possibility of certain relatively practical or political interventions, but because those interventions and even the science "itself" are *acts of the living*, they cannot help but displace the very truth that would count as scientific and its object or objectivity. There are only sciences *of* life, double genitive. This circular structure suggests a hyperbolic form of responsibility by which we must hold ourselves accountable even for what will count as truth. Whatever anxieties this might expose, given that it seems to offer a sort of opening to "science denialism," it must be recognized that it is only by means of this same opening that a sort of science-affirmation receives its first chance or risk. It may always be necessary, even to act in the name of science and something like objectivity, that science or what has gone by that name be denied, that certain of its established authorities, individual scientists, credentialing and research-funding institutions and governments, publications and professional societies, the media and other sources for communication of its results, and so on be placed in question. No authority can ever guarantee, by virtue of its authority or sanction alone, that it will remain on the side of truth or justice in its operations. Science is inimical to the authoritarianism that would simply imagine an establishment somewhere else that was responsible for validating the truth, and thus there is a possibility of science at all only there where its very possibility is radically placed in question, where we are exposed to the risk that nothing worthy of the name has ever yet transpired.

Viral Rhetoric, Viral Truth

There will never be a *pure* virus. Its lineage or generations, its concept and schema, will always be the host or breeding ground of some alterity. Even if we take it to be the concept or figure of *impurity* "itself," such impurity cannot cut a figure in intuition, form a family tree, or take on rigorous definition without compromising itself with what it is not — purity. As the most impure impurity, it even welcomes and allows the pure to breed inside it. It will never be as simple as pointing to a certain phenomenon or material structure within experience, or offering a definitional formula, to know what "the virus" is. Nonetheless, even if one can promise or threaten this in advance, and demonstrate it according to the best logic or intuition, that will never have stopped the virus, word or thing, from simulating and dissimulating every possible *effect* of conceptuality, generativity, and family resemblance. It will always be *as if* "virus" were a coherent concept capable of rigorous logical relations with all its species, genera, and contrast classes, *as if* it formed a family of descent, of generational, reproductive, and filial relations, and *as if* its members shared an intuitive resemblance. This *as if*, which lets what is not breed within what is, and makes the intuition and realization of every concept and perception possible while nonetheless compromising it in principle — it is *as if* "as if" were a sort of parasite or virus itself.

The virus is not unique in deconstructing the borders that would be necessary for its existence — if anything, defining the virus as a distinct object is an attempt to contain this force of destabilization. On several occasions, Derrida observed a continuity between the operations of the virus and those of deconstruction. In "The Rhetoric of Drugs," an interview he gave in April of 1989, he discusses how HIV has affected our sense of subjectivity and intersubjectivity:

> [L]et us limit ourselves in any case to this fact of our time that I believe to be absolutely original and indelible: the appearance of AIDS. This is not just an event with immeasurable effect on humanity, both on the world's surface and within the experience of the social bond. The various forms of this deadly contagion, its spatial and temporal dimensions deprive us henceforth of everything that a relation to the other, and first of all desire, could invent to protect the integrity and thus the inalienable identity of anything like a subject: in its "body," of course, but also in its entire symbolic organization, the ego and the unconscious, the *subject* in its separation and in its absolute secret. The virus (which belongs neither to life nor to death) may *always already* have broken into any "intersubjective" trajectory. [. . .] And at the heart of that which would preserve itself as a dual intersubjectivity it inscribes the mortal and indestructible trace of the third party[.][33] (Derrida 251)

Derrida's reflections on how virality transforms our relationship to others, and thus our relationship to ourselves, are applicable in full to the transformations of the public sphere that have taken place during the global COVID-19 pandemic. The virus reveals that some alterity, some virality, has always already been at work in shaping the very idea of a subject and a community. Every time self and other gather, some third is present in absence, not being simply another self like the others, but remaining as the specter of subjectivity without which this encounter could not take place. If it is possible for various forms of prophylactic and virtuality to become the norm and enabling connectivity of this community, for it to become a given that condoms, PrEP, masks, or video chats are to be preconditions of certain forms of social gathering, it must have always been the case that something like a virus, something neither simply human nor living yet inseparable (in a fashion that remains to be explored) from our nature and culture, has always been at work in making this sociality possible. The borders of public and private shift in ways both subtle and massive, which reveals that there was never simply a public — the true limit was never simply known and publicizable — yet there is certainly no privacy that would straightforwardly belong to oneself or one's intimates. The virus eavesdrops; one never knows what it knows. Every convention and contract, including rituals as quotidian as a handshake or as intimate (in some cases) as sex — both of which can occur in relatively private or public contexts — are capable of being transformed by a virus (a *viral risk*, never simply absent or present) and thus have always been countersigned by some alterity. As has always been the case, one who chooses not to keep pace with these social norms, which have never been a fixed or natural given, risks cutting themselves off all the more from the supposedly originary community they hope to preserve — though one may always form a community of dissent in the process.

The self and the others, the immunity and community that we hope to preserve by these detours, have always been subject to this structure, not simply of returning to themselves from an other, but returning from an encounter with an other not at one with itself, with the life-death of a viral alterity. The self is a virus to itself. That certainly does not prevent us from identifying certain pathogens as threats and working to control their dissemination, but it will pose certain difficulties if we hope it will be possible to define *the* virus and to distinguish ourselves from it. Precisely because its effects have always already been in play, and are the very ground from which anything like an individual inherits its possibility and impossibility, there will never be a simple opposition between life and virus or autonomy and parasitism. For this reason, Derrida repeats that what has changed, as these undeniable and far-reaching displacements spread globally in the face of certain pandemics, is not exactly the difference between vitality and virality (which is not a simple difference), but a transformed *legibility* of the role virality plays in countersigning the social contract:

> You may say this is how it's always been, and I believe it. But now, exactly as if it were a painting or a giant movie screen, AIDS provides an available, daily, massive *readability* to that which the canonical discourses we mentioned above [discourses for or against legal drug use, both of which invoke some value of the natural or the self] had to deny, which in truth they are destined to deny, founded as they are by this very denial. If I spoke a moment ago of an event and of indestructibility, it is because already, at the dawn of this very new and ever so ancient thing, we know that, even should humanity some day come to control the virus (it will take at least a generation), still, even in the most unconscious symbolic zones, the traumatism has irreversibly affected our experience of desire and what we blithely call intersubjectivity, the relation to the alter-ego, and so forth [. . .]. As sudden and overwhelming as it may be, this event had heralded itself even before we could talk about history or memory. The virus has no age. ("Rhetoric" 251, 254).

No virus or parasite would be possible without some virality that was there already, shaping the self-relation of the living. This virality is no more simply present today, as this or that contagion or pandemic, than it is or ever has been simply absent. Virality belongs to nothing, and nothing belongs to it — it obeys only the troubling logic of *alterity as such* or *pure parasitism,* which is capable of occupying every form, even the most apparently autonomous, without ever simply being anything. For this reason, I would not attempt to claim, now or ever, that we have been living in the Age of the Virus. Even if what goes by the name *virus* today is often counted to be the most abundant form of organic matter, to the point of creating effects at a geological scale (Suttle; López-García et al.), I would not say that we inhabit a Virocene. At the very least, it is necessary to recognize that even the name *virus* may prove insufficient, too rigid or self-identical for the forms of self-subversion or subterfuge that a *virality without figure* will undergo. If this virality is not simply a positive term, which may or may not befall the living in the course of their life cycles, but an alterity that is always already at work in co-constituting or de-constituting the living, then it becomes quite difficult to say what the *literal* virus is. Anything that elicits or displaces these relations of self and other has the structure of a virus. Derrida introduces the figure of the computer virus to pose this question:

> For example: in the case of computers, is the use of the word "virus" simply a metaphor? [. . .] The *prerequisite* to this sort of problematic would have to concern rhetoric itself, as a parasitic or viral structure: originarily and in general. Whether viewed from up close or from far away, does not everything that comes to affect the proper or the literal have the form of a virus (neither alive nor dead, neither human nor "reappropriable by the proper of man," nor generally subjectivable)? And doesn't rhetoric always obey a logic of parasitism? Or rather, doesn't the parasite logically and normally disrupt logic? If rhetoric is viral or parasitic [. . .] how could we wonder about the rhetorical drift of words like "virus," "parasite," and so forth? ("Rhetoric" 472n9)

Any hope that we could fix the true limits of the virus — for instance, that we know absolutely the difference between a literal biological virus and a computer virus, and thus could never be taken by surprise by a computer that could "resurrect" a virus, a virus that was never dead because it was never quite living — any such hope founders against this necessity. In order to fix the literal and figurative valences of virality, it would be necessary for truth and rhetoric to be domains larger than virality, within which its effects could be circumscribed. But now it seems, not that we will be able to sort all kinds of viruses into literal and figurative, *but that truth and fiction are kinds of viruses.* One should not treat this as a reversal, as though we could then articulate an absolute or fundamental virology — as Derrida says, we are rather preoccupied by what "logically . . . disrupts logic," what is neither simply outside nor within. The virus is not a simple truth because it *is* not, it is nothing simply present and has never belonged to the lighting or clearing of being — rather, it is *as if* the "proper" were only a temporary formation of viral replication.

The logic of rhetoric would suggest that a *figurative* virus, such as we might take a computer virus or rhetorical virus to be, would be something less familiar than the *literal* virus, which was invoked precisely to make the abstract intuitive. However, if literalness and rhetoricity are only intelligible in relations of parasitism, then there is a certain virality that is more familiar than familiarity itself — closer to us than anything that could count as our own or our essence. What we tend to think of as a literal virus — which we might try to designate as biological, though it is often thought not to be alive — this virus is no closer to the truth or essence of virality than any other, for instance a computer virus. Each becomes what it is only by occupying some conceptual or iterable terrain — every virus is a virus of the virus, a meta-virus.[34] There is nothing more essential to virality than this, that the viral is what first makes the proper boundaries of life and the literal intelligible by displacing and corrupting them. A computer virus has the potential to displace every border that we witnessed a moment ago being destabilized by a "biological" virus — the boundaries of communities and nations, the exchanges of goods, information, energy, and so on that take place across them, systems of health care and governance, the agencies of security and defense that are supposed to manage and preserve these frontiers, all must be rethought in light of these viral possibilities. The very meaning of public and private (including where to locate the limit of the body proper, a life of one's own, the biological and natural) is in question if even the most secretive or encrypted code can be found out by some virus, and if even the most apparently public publicity may always hide this unknown viral threat.

It only requires bringing to mind everything Wimmer's experiments revealed about the nature of life and virality to recognize that the boundaries of life, species, genetics — everything "biological" is subject to displacement by a computer virus that is supposedly outside of the literal or biological domain. Our repositories of knowledge of the living, such as those public databases that contain the genetic sequences of viruses, and all of the experimentation, synthesis, and technology that depend on them — including the maintenance of our own life and health — are at risk, at least in principle, of being corrupted by a computer virus. It is all the more difficult to draw a firm line between these domains as our lives become increasingly mediated and virtualized, a trend that has been particularly visible during the pandemic, though it has always been underway. At the very least, the form in which this information is stored and disseminated must take into account viral risks, must transform itself in advance to anticipate and evade them.

To give just one example of this principle in action, not of a computer virus per se, but of a *bug*, one recent study estimated that about twenty percent of research in genetics may have been distorted because Microsoft Excel mistakenly formatted the names of certain genes as dates (Zeeberg et al.; Ziemann et al.; Ingraham). This bug has distorted the very science by which we understand our "own" life and genetics — and in order to correct the error it was easier to change the name of those genes than to fix our technology (Bruford et al.; Vincent). Such an interface, where the supposedly literal and figurative, biological and technological, overlap, confuse each other and themselves, do the work of the other by coming closer to its "self" than even it does — such a virus of the virus will always in principle be possible. We would probably seek to distinguish virus from bug by invoking certain modalities of agency and intention (a bug is the programmer's accident, a virus is another programmer's intention or counter-purpose), but *virality* is precisely what allows for the stabilization of an intentional formation only on the basis of the "same" forces that let it be carried away. There is some virus anywhere reading and writing take place, which is to say, wherever *misreading* or *counter-reading* is possible.

Aspects of the form this interface takes in the present, and even its intelligibility to us as the confusion of forces or tongues that ought to remain separate and separable, depend on contemporary informational and linguistic technologies (e.g., computers and words) that are anything but timeless. Nonetheless, it could be shown that something like this confusion has always been possible. There has never been a life without such borders and territories, and thus never a life that has not been a trace to "itself." As such, every relationship, including self-relations, has been exposed from before the beginning to the sort of miscommunication, espionage, and subterfuge that today we call *virus*. If this language seems too human to capture the operations of the virus "itself," one need only remember that the language a "biological" virus can read and rewrite is figured as a *genetic code*, and an *alphabet* whose basic operations are *transcription* and *translation*.

Derrida returns to the themes of his "The Rhetoric of Drugs" interview on two occasions, in both cases suggesting a kind of kinship between deconstruction and virality. In "Circumfession," he mentions rather briefly and elliptically that "the virus will have been the only object of my work," (Derrida 91–92), and elaborates in another interview:

> I often tell myself, and I must have written it somewhere — I am sure I wrote it somewhere — that all I have done, to summarize it very reductively, is dominated by the thought of a virus, what could be called a parasitology, a virology, the virus being many things. I have written about this in a recent text on drugs. The virus is in part a parasite that destroys, that introduces disorder into communication. Even from the biological standpoint, this is what happens with a virus; it derails a mechanism of the communicational type, its coding and decoding. On the other hand, it is something that is neither living nor nonliving; the virus is not a microbe. And if you follow these two threads, that of a parasite which disrupts destination from the communicative point of view — disrupting writing, inscription, and the

> coding and decoding of inscription — and which on the other hand is neither alive nor dead, you have the matrix of all that I have done since I began writing. (Derrida et al. 12)

My goal in focusing on the virus and virality is to open thought to deconstruction, to the displacement of the self-certainties of life and the life sciences.

Life has been thought as reproduction, species as reproductive community, and even biological science as a certain *reproducibility* of our knowledge of the living. This reproduction or reproducibility is the opening of a structure to what it is not or not yet — it receives its chance and its risk from what threatens to subvert it, a virality or viro-graphy, an iterability that promises life and the sciences of life will remain a practice of deconstruction.

Part I

Of Virology

Chapter 1

Trees of Life

This chapter deals with three phenomena that complicate the foundation of evolutionary science. This foundation is the reconstruction of phylogeny, *the relationships of descent among species from the origin of life to the present. Phylogeny typically takes the form of a* Tree of Life, *whose branches represent the origin of new species from a common ancestor.*

One process that disturbs the reconstruction of phylogeny is Horizontal Gene Transfer (HGT). *While organisms receive a set of genes at their conception, it is possible for many organisms to continue to receive genes throughout life from unrelated or distantly related individuals. This is especially true of* prokaryotic *cells (such as bacteria), which are defined by their lack of internal structure relative to* eukaryotic *cells (which comprise a subset of single-celled organisms, as well as all fungi, plants, and animals). If one represents these "horizontal" gene transfers as lines crisscrossing the branches of a tree, then the "tree" of life begins to resemble a* web *or* network.

On the other hand, it has been recognized by biologists themselves that if the relations of inheritance among the living do not originally take a tree-form (if the constitution of a "species" depends to a certain extent on "horizontal" transmission), then it is not possible to speak of their "horizontality" either. "Horizontal" simply means the lateral crossing of the branches of a tree, and any tree-pattern is fictive if "horizontal" transfers have been sufficiently prevalent. What becomes of the basic tasks of evolutionary science and our concepts of heredity or descent if both "vertical" and "horizontal" or "tree" and "web" are equally fictive representations?

Another concept haunting the reconstruction of phylogeny is what biologists call analogy. *Two structures or genes are said to be* homologous *if they are related by descent from a common ancestor. On the other hand, they are analogous if they are similar in appearance or function but not because of shared ancestry (for example, if they adapted separately to similar environmental pressures — also called* homoplasy*). The history of these concepts reveals an undecidability of typological/structural/essentialist/idealist and evolutionary/historical accounts of life.*

The third category to be considered is that of loss. *The possibility that an element, including a gene, can be lost in the course of descent complicates the reconstruction of phylogenies. It is possible that the complete absence of a trait is nonetheless a mark of common ancestry.*

At the end of the chapter, I introduce a term, gone, *to deploy what Derrida referred to as a "double science" or "bifurcated writing," which would overturn a tradition while nonetheless inevitably belonging to it (*Positions *41–42). The deconstruction of these apparent oppositions: vertical/horizontal, homology/analogy, inheritance/loss, reveals them to be* undecidables. *That is, it is not possible to treat them as positions, to claim in more than a provisional sense that one prevails over the other in determining the history or essence of life. In a certain sense, gone belongs to the linguistic, philosophical, and scientific traditions under consideration here but disturbs the functioning of their oppositional logic, by bringing together absence and generativity (as I will explain). It suggests an opening by which this tradition may be inherited otherwise.*

***

The concept of a parasite or symbiont, and thus the concept of virality, implies exteriority to the natural family, to the "vertical" chain of relations secured by "natural" reproduction. (It is not typical to think of a baby in the womb as a parasite, even if the demands it places on its gestator are similar.) The image of life in which these vertical relations are predominant is that of a family tree, or of all the living assigned a place within a vast Tree of Life. The virus not only traverses these branches, but tangles or anastomoses them — it confuses and is confused with properly familial and reproductive relationships by making itself an agent of heredity, or simply by making itself indistinguishable from one.[1] In short, given that this infectious intrusion is capable of transforming the heredity of its host, it begins to resemble an ersatz ancestor. Is it possible for a parasite, a horizontal displacement, to be more fundamental than the vertical lines it crosses? There is no horizontality to speak of if there is not something like a tree already standing, yet the very soil from which that tree grows is formed of relations that are no more purely vertical than horizontal, a sort of bend sinister that renders even illegitimacy illegitimate.

The Tree of Life grows at the crossroads of nature and logic. It depicts the relationships of living things according to their supposed natural birth — representing generations of parent and offspring, or the origin of species, or the higher orders of taxonomy, branching off from their ancestors in a fashion that is meant to portray their historical relations of descent. At the same time, in a fashion one hopes will be coherent and nonconflictual, it depicts logical relations of *genus and specific difference*, the dichotomous branching of taxonomy. The superimposition of phylogeny and taxonomy has been an aspiration of biologists at least since Darwin, though predecessors could also be found — this tree represents what, like the form of the good, would not only be the natural nurturance of its members but also that by which they are known (Plato, *Republic* 509b). The tree's dual significance, representing both reproduction and knowledge, will not be its only resonance with the trees of the story of Genesis, and the semantic range biblical translation gives to the verb *to know*.

Darwin's deployment of the tree-figure, which he acknowledges as a borrowing or transplant, attempts to superimpose the trees of life and knowledge. As if a single tree would limn both the generations of life and their logical relations, diachrony and synchrony, genealogy or phylogeny and taxonomy:

> The affinities of all the beings of the same class have sometimes been represented by a great tree. I believe this simile largely speaks the truth. The green and budding twigs may represent existing species; and those produced during each former year may represent the long succession of extinct species. At each period of growth all the growing twigs have tried to branch out on all sides, and to overtop and kill the surrounding twigs and branches, in the same manner as species and groups of species have tried to overmaster other species in *the great battle for life*. The limbs divided into great branches, and these into lesser and lesser branches, were themselves once, when the tree was small, budding twigs; and this connexion of the former and present buds by ramifying branches *may well represent the classification of all extinct and living species in groups subordinate to groups*. Of the many twigs which flourished when the tree was a mere bush, only two or three, now grown into great branches, yet survive and bear all the other branches; so with the species which lived during long-past geological periods, very few now have living and modified descendants. From the first growth of the tree, many a limb and branch has decayed and dropped off; and these lost branches of various sizes may represent those whole orders, families, and genera which have now no living representatives, and which are known to us only from having been found in a fossil state. [. . .] As buds give rise by growth to fresh buds, and these, if vigorous, branch out and overtop on all sides many a feebler branch, so by generation I believe it has been with the great Tree of Life, which fills with its dead and broken branches the crust of the earth, and covers the surface with its ever branching and beautiful ramifications. (Darwin, "Variorum" 1:129–30; emphasis added)

A single branching pattern would capture both "the great battle for life" and "the classification of all extinct and living species in groups subordinate to groups," descent and taxonomy. Multiple figurations and representations are vying to "overmaster" each other in this "simile." On the one hand, if the tree represents the survival and descent of natural organisms, only its "green and budding twigs" would be alive at present. On the other hand, as a logical structure, representing the "classification" of life, it sublates this history into a sort of eternal life. It is this blinking alternation between the tree as a film or crust of life over an orb of death, or as vital down to its roots, that led Darwin, in a notebook from two decades before the *Origin*, to suggest an alternative figure, the coral of life: "The tree of life should perhaps be called the coral of life, base of branches dead; so that passages cannot be seen" (*Notebook B* 25). Moreover, Darwin pictures only those lineages with living descendants as still attached to the trunk of this tree. Extinct lineages are branches that have died and fallen to earth, a metaphor Darwin extends and spatializes to account for the geological strata of the fossil record. The Tree of Life "fills with its dead and broken branches the crust of the earth" while "covering the surface" with its living "ramifications." This figure would be apt *only if all fossils were from extinct lineages*.[2] A fossil from an extant lineage now suggests a branch growing from deep within the earth and breaking out over the surface. Noting these incongruities is not merely to suggest some rhetorical or poetical deficiency unrelated to Darwin's biological subject matter. Rather, as we will see, the undecidabilities haunting any figuration of life or inheritance possess the work of the scientist, and even make it descend from a poetizing vocation.

For Darwin, the logical or taxonomic aspect of this tree is not simply synchronic. He pictures a generative logic, one that unfolds as nature itself, bringing forth the dependent categories from the higher in a process of becoming. Darwin presents his tree as an illustration or perhaps justification of this claim, that his theory of descent with modification (what we call *evolution*) explained, in a word, the *nature of logic*, explained why nature took a logical form and why logic was to be found in nature:

> It is a truly wonderful fact — the wonder of which we are apt to overlook from familiarity — that all animals and all plants throughout all time and space should be related to each other in group subordinate to group, in the manner which we everywhere behold — namely, varieties of the same species most closely related together, species of the same genus less closely and unequally related together, forming sections and sub-genera, species of distinct genera much less closely related, and genera related in different degrees, forming sub-families, families, orders, sub-classes, and classes. The several subordinate groups in any class cannot be ranked in a single file, but seem rather to be clustered round points, and these round other points, and so on in almost endless cycles. On the view that each species has been independently created, I can see no explanation of this great fact in the classification of all organic beings; but, to the best of my judgment, it is explained through inheritance and the complex action of natural selection, entailing extinction and divergence of character, as we have seen illustrated in the diagram. ("Variorum" 1:128–29)

Logical or conceptual definition, like life "itself," also takes a branching form: many species radiating from a single genus, many genera from a family, and so on. Darwin is arguing that this pattern can be explained, and explained as natural, if its cause lies in the similarly branching descent of reproductive relationships. A natural logic, a natural logos and *legein* — which we could very well translate as selection — would be more than a happy coincidence or preestablished harmony if we could find its cause in nature itself. Yet, if nature does not exist until origin and essence can be recovered or invented by borrowing a logical form, we the living will be turning in a circle that makes every tree a kind of transplant or parasite. It would be premature to explain away our sense of wonder.

The tree's naturality runs deeper than any claim that it simply mirrors or represents relationships of reproduction as they once existed in nature. There would be no nature to speak of if it could not be made to fit within such a tree. Nature suggests the value of the innate or inborn — in order to discover the proper possibility that belongs to a thing itself, its nature or essence, its present state must be analyzable into inner accomplishments and those accidents or externalities that have attached to it like so many parasites or viruses. Nature can only be recovered if it can be contrasted to non-nature, whether that is some kind of culture and cultivation, or a squandering dissemination, nature's deviation and displacement. This recuperation is only possible, there was only ever a nature to begin with, if it was there from the origin. There must have been a proper origin, the one we call conception or birth, that imbued the natural thing with its proper possibilities, that were then either accomplished or stymied. We only ever discover nature in various degrees of corruption and perversion, as the imbrication of this "proper" origin and possibility with an undecidable fabric of distortion or displacement — a science that pretends it can restore this origin that never was is just as much the inheritance as the creation of this "nature." It naturalizes by narration, a fictional or virtual pretense of origin that we could call *naturration*.

The science of genetics operates by naturration. On the one hand, this is legible in the dominant form evolutionary theory developed throughout the twentieth century, and which is still taught to students of biology today, one that privileges inheritance by conception over the "inheritance of acquired traits." On the other hand, even those emerging challenges to this orthodoxy can only displace its concept of origin by re-inscribing the powers of genesis in a no less fictive fashion. Still, if there has been a tendency to tie everything in the form, function, and behavior of an organism to innate possibilities endowed by a material substance (the genes) received at the moment of conception, this suggests a striving for the purity and simplicity of origins. This genotype serves the demand of naturration because it is understood to be present from the origin. There is an origin of *what is* only to the extent that whatever presents itself in the present can be tied back to a reason, ground, or cause understood as fixed and innate. Even if it were possible for fixity to be suspended, for possibility to change into impossibility and vice versa, the task of the scientist or metaphysician seeking the truth would simply be to seek out the law of those changes, which would be nothing other than the true law of the possible. Birth is birth and nature is nature, the origin, only to the extent that the *possibility of what will be* can be attributed to it. One can ask *why?* and receive one's answer: birth.[3]

Now, life or anything else will only be fit within arborescent representation by invoking such an origin — by narrating or naturrating its history as a series of births, each depending on the possibility endowed by the one preceding. One or more branches stand as forking paths from a trunk representing the relations of a generation of children to the generation of their parents. It will only be possible, on such a diagram or map, to represent the generation of grandchildren if the same beings who once were those children accomplish their own reproduction by means of a possibility granted by their birth, their parentage. If something had to come along in the course of this generation's life, in the growth of its green and budding twigs, if some stranger, guest, parasite, or disguised god had to befall it and make the impossible possible in order to account for its subsequent history or life story — even the chance of such an encounter uproots our fragile tree.

A common pest described by the theorists who prune this tree is Horizontal Gene Transfer (HGT), which is almost a *contradictio in terminis* — if we still think of the "gene" as a unit of heredity. Horizontal Gene Transfer names a process by which genes can be traded amongst organisms (including between a virus, which we do not yet know whether to categorize as living, and its host). The "horizontal" in this term refers to the lateral movement between branches of an implicit Tree of Life — it is the contrast class to those "vertical" relationships by which a complement of "genes" is received through natal inheritance, natural birth.[4] Eugene Koonin, to whom we'll return in a moment, defines HGT as "the transfer of genes between distinct organisms by means other than vertical transmission of replicated chromosomes during cell division" (*Logic* 119). A horizontal transfer implies that an organism already born receives a "gene" from a source that more or less disregards vertical relationships of consanguinity (though all life, we are led to believe, is at least distantly related). A phylogenist then faces an aporia; either they can attempt to represent every such chance encounter as a birth, as a tributary to hereditary power, or one is left with a lineage of descent that cannot account for the supposed repository of the possibility and actuality of the life of the living, its "genetics."[5]

The question of whether this transforms or mutates life's tree is not simply a matter of degree, of "how" horizontal our relations are, such that the problem could be posed as a simple alternative left to the decision of the informed investigator. As soon as this horizontality is possible, always already, there is no simple sequence of historical facts (representable by the linearity of a tree) that would then come to be tangled or obscured by horizontal happenstance. For instance, it is entirely possible that the "child" of a "parent," even a genetic clone formed by binary fission, might pick up genes by this horizontal process then pass them to a cell of the parental generation. This involution, analogous to incest, is not simply representable as a graft or inarching on our supposed tree.

The child is now the parent of its parents, the grandparent of its future siblings, and the disorder thus introduced will ripple across every relation past and future in a fashion that is no longer representable or schematizable by any organic or artificial figure. Even if we assigned it an arbitrary X or a distinct mark in our diagrams, there would be no way to analyze the significance of such a mark without resuming the deconstruction of the entire arborescent project.

Nothing is more everyday than this deconstruction of life. It is the most commonplace thing, and one could even surmise that there would be no life without it. There could be nothing at all. That is, it is not *only* "life" that transgresses origin and essence to stand in the present as a haunting trace — one should not attempt to ground this deconstruction or denaturation in some vital creative force that would grant life and life alone a power beyond possibility of self-creation through self-abnegation. Rather, there is no "present" to speak of, no living and non-living, natural and non-natural things, until something like essence and origin have begun to be titrated from their dissolution in a field that would have neither subject nor object, past nor present, truth nor error, without them. The "origin" of the origin, prior to any imagined natural or vital power, is the derivative and disseminated trace in which vertical and horizontal, the proper and its corruption, essence and accident, are undecidable.

Nor can "horizontality" be limited to genetic processes or the organisms thought to be subject to them to the greatest degree. Even some of those scientists who advocate for a radical restructuring of the representation of life on the basis of Horizontal Gene Transfer limit this reorganization to the kingdoms of life in which it is most active, primarily the prokaryotes, with perhaps some eukaryotes included, but certainly not animals, and thus not human beings, nor the scientists forming these classifications.[6] What is being unearthed is not a historical occurrence that might or might not have happened (such as a horizontal transfer), but an uprootedness that co-constitutes the possibility of the "origin." In fact, the name *horizontality* no longer suits it given that there is no verticality prior to it. Perhaps we could call this *a-filiation*, which is *both* the uprooting by which the impossible at every moment intrudes on possibility and actuality, rendering cause and effect or parentage and generation illegitimate, depriving them of filiation, and yet *also* the chance granted to unanticipated and unthinkable *affiliations* unbound by natural kind or kin, even first forming that nature by perverting it. Even if we understand this a-filiation as somehow fundamental, not exactly as primary but as having nothing prior to it, derivative yet underived, it is nonetheless not opposed to the taxonomies or classifications that appear artificial and arbitrary in its borrowed light. There is only a-filiation where fictive kinship structures, or logical and causal orders of relation, have been laid down in advance, even if a-filiation is only the overthrowing of every such positing or structuration. This a-filiation is at work or in play forming and unforming every grouping of the living and non-living, troubling their borders.

A particular horizontal process, such as Horizontal Gene Transfer, may be relatively restricted in a certain lineage, but horizontality as a-filiation can never simply be excluded. In fact, it is most "powerful" or uncanny in that its disjunctive haunting draws from a certain absence or alterity, which makes the presence of the present shine with a borrowed luster. It will not be possible, within these limits, to fully explore epigenetics, symbiogenesis, developmental systems theory, plasticity, eco-evo-devo, niche construction theory, and every other discourse that today is placing in question the simple representations of the gene and population genetics.[7] Where it was once thought that inheritance passed entirely from parent or parents to offspring at the moment of conception, today it is increasingly recognized that heredity depends on chance encounters with the biotic and abiotic environment, well beyond the "natural" family. The "gene" has never simply been a material substance, but, as a hypothesized unit of inheritance, receives its identity from the cross-generational resemblances it can be correlated with, and these resemblances are seen to depend on "externalities" outside the nucleus. The attempt to treat this irruption as merely a broader network of possibility, to bring the whole world within a developmental system or extended phenotype, can only domesticate what continues to destabilize these and any systems of inheritance. The openness of this system on a future that has never finished arriving necessitates that the *impossible* can overtake it or sur-vive it at any moment, an impossible that co-constitutes all possibility. This necessary impossibility is the "horizontality" of the origin. What will count as the vertical and proper descends or a-filiates from what has never simply existed.

This non-unified and unbounded field is the one faced by those we call the living, as well as the scientific investigator, whose work as a *geneticist* is not simply one task among others. The scientist has no object of study, no nature to begin with, unless vertical can be separated from horizontal, even if the processes so named render the very distinction untenable. But the world faced by the scientific investigator is both and neither at once, in an impossible simultaneity. Nor is this scientific investigator a special character whose work we may choose to take interest in or not, but represents each and every one of the living whether we operate as "folk" or expert biologists. Every living thing operates as a folk biologist as it forms its alliances and disalliances, sorting friend from foe and risking and reappropriating species-being by choosing or shunning its mates. These cultivated relations take the form of treelike categorizations, a system of logic and life that is as much inherited as invented.

Whether we are naïvely or intuitively classifying perceived resemblances, or opening a cell to itemize its genetic contents, any classification or definition remains unstable if it tries to categorize whether elements have arrived through vertical filiation or some horizontal process of borrowing, imitation, intrusion, infection, gift, and so on. *It is not even possible to rigorously determine how to give meaning to these categories*, what should count as a birth or as the innate, what is true and legitimate filiation, and what represents an adventitious externality befalling it. The problem is (at least) twofold, involuted and redoubled with the forces of dispersion and dissemination that it gathers: (1) For one who hopes to decipher the family or phylogenetic relations of the living, the very possibility of horizontality is a constant stumbling block — anything that can be observed in the present, say, a gene or trait possessed by two different organisms, may be a mark identifying that they have descended from a common ancestor who endowed both their lineages with this heredity, or one may have passed this mark to the other just yesterday, having only borrowed it from another the day before. Anything that can be traded horizontally, by imitation or infection, can confuse the traces of family history or form alliances that break with reproductive genealogies. (2) Yet, on the other hand, this "same" possibility is the source of its own impossibility. One should not speak as if there were a given history of vertical or horizontal relations, which are merely obscured by the similar effects or effect of similarity they produce. It is never possible to ultimately decide, for any event in the past, present, or future, whether it has simply been vertical or horizontal — what should count as normal inheritance or birthright as opposed to gift or infection, what as innate and what as acquisition, what as kin and what as unrelated could only form a secure distinction if there had ever been the sort of history that is deconstructed by these undecidable exchanges.

This element here, say, this gene, may have reproduced faithfully throughout an ancient lineage, or may have arrived yesterday by a "horizontal" transfer from a distantly related cell or virus — yet, it is not simply a separate or separable question to ask *what should count* as reproduction, as relation, as vertical lineage or its horizontal transgression; in fact, the science or process of inheritance we call genetics can never ground a history without a typically unacknowledged supplement. The gene is so far from telling us what life, reproduction, and inheritance are, that it is not even identifiable unless these terms have been fixed independently of it. Why, for instance, would the calculations of a population geneticist tend to count or account for the presence of a gene in two "organisms," but not the millions of copies in each organism's individual cells? In short, because the question of life and of what it means for life to reproduce has been presupposed by this investigator — the "reproduction" of the gene depends on this preexisting theory or lineage, rather than grounding it. Thus, the "two" questions of verticality and horizontality tremble together, they come too close to simply be delineated, each implies an answer to the other while also fictionalizing or naturrating itself in the process. Science "itself" is in the a-filiative or incestuous position of needing to borrow from that for which it is supposed to provide the ground and origin — it draws its resources from its own effects.

Analogy and Homology

Neither same nor different: the question of analogy. The investigator observes a similarity, a resemblance. This may be observed between two "genes" or between two "phenotypic" traits. Nothing tells us whether this resemblance exists by essence or by existence — in evolutionary terms, this is interpreted as meaning either that this similarity exists because two organisms have descended from a common ancestor (homology), or that two separate lineages have happened to independently evolve a similar form (analogy). Nothing in principle limits this question of analogy, of "mere" similarity, to this evolutionary framework — Darwin's typological predecessor Richard Owen and even Plato and Aristotle required an analogous distinction in order to sort essence from appearance. The question of analogy is pressing whether its opposite is understood genealogically or typologically.

It is possible to identify historical precursors to contemporary debates concerning the nature of homology. It remains to ask, then, what is the true nature of the relations among these terms or structures of being and thought, whether their relationship is one of historical inheritance and mutation, or of a structural and essential necessity that finds itself everywhere confirmed. If this question of the homology of homology is not just possible but necessary in order for us to know the first thing about what we mean when we deploy this term, then its definition will not be a simple matter — it will always be some alterity that grants the possibility and impossibility of what can no longer be contained by any single name.

It would not be possible to offer an exhaustive history of the concept here, but passing through certain examples of its aporetics can displace any security we might feel regarding the historico-evolutionary form the concept often takes in the present. Homology is sometimes traced back to the work of Aristotle, where we find one of the first projects known to history of the taxonomy of living things (*HA*; Gould, "Evolution" 66; Hall 4). Plato provides another important precursor, in that he attempted a dichotomous tree-classification of the living (*Statesman* 261b–6d). Plato's method of diaresis, dividing higher forms into subspecies, is perhaps more like certain contemporary biologists' Trees of Life than anything in Aristotle, in that it seems to aspire to create a single branching structure for everything. One can dissect a form such as life (*ta empsucha*), for example, to arrive at the human being as the featherless, two-legged animal.[8] Here, where there is no pretense of discovering historical relations of descent among species, the tree-structure is no less necessary in order to arrive at an essential categorization. Though a homology/analogy distinction does not appear explicitly there, it is implicit wherever Plato or anyone else distinguishes true from false categorical belonging, and this necessarily circular distinction of essence from appearance is the shared ground of essentialist and evolutionary projects of taxonomy. As much as modern biology likes to imagine itself as having advanced beyond "idealist" Platonic premises, the historical or genealogical approach is in truth no less essentialist, given that its narrative depends on positing such categories. Aristotle was explicitly critical of the totalizing pretense of Plato's diaresis, and many of the arguments he leveled against it recur in debates among contemporary systematists (Aristotle, *PA* I.2,

4.642b–644b; Panchen, *Classification* 110–11). While no term exactly parallel to *homology* appears in Aristotle's works, his role as a precursor or ancestor is acknowledged because he does distinguish true generic belonging from relatedness by analogy [*kat'analogian*] (M. Wilson). Animals belonging to a common *eidos* [species or form] would be identical, and several such species can be classed under a common *genos* [genus or lineage]. Generic belonging is distinguished from correspondence by analogy, so that a genus such as *bird* [*ornithos genos*] can incorporate species with larger and smaller wings, but any wing is only related by analogy to the forelimbs of another genus. Now, Aristotle does not imagine that the genera could be categorized under still higher categories, such as those from which Plato (or our contemporaries) attempt to derive them. They rather originate from what we would call *folk biology*: "The proper course is to endeavor to take the animals according to their groups, following the lead of the bulk of mankind [*hoi polloi*], who have marked off [for instance] the group of Birds and the group of Fishes" (Aristotle, *PA* I.3, 643b10–12). Aristotle constructs something like a series of bushes, a hedgerow rather than a tree, whose highest points are arbitrary and disconnected. Analogy forms the links between these structures, and within them something similar to homology (again, without being grounded in descent but descending from the essence) defines generic belonging. Yet, we can see that outside the system a sort of transcendental analogy-homology grants its provisional stability and instability. The folk categories (fish, birds, etc.) are themselves formed out of intuitive resemblances (analogies) and/or linguistic and cultural inheritance (homology).

Given the evolutionary form these categories (homology and analogy) often take today, one might expect that we have inherited them from Darwin. Curiously, while these concepts are certainly found there, they are received by Darwin as an inheritance from an unlikely source. Prior to Darwin's use of these concepts, they could be found in the work of Richard Owen.[9] Owen was Darwin's contemporary and has been somewhat of a fall guy for whiggish representations of the pre-Darwinian history of biology (Mayr, *Growth*).[10] In Owen's work, distinguishing homology from analogy was not a matter of recovering phylogeny (the true lineage of descent), but of identifying relationships whose essentiality was grounded in an archetypal or typological vision of life (Panchen, "Richard Owen"). Declaring two vertebrae or limbs homologous did not suggest per se that they descended from a common ancestor, but that they belonged to two species or tokens of a single archetype. Owen spoke of Plato and the forms when describing these archetypes, and even their invention in the mind of the creator, which gave his work a decidedly non-modern cast, and made him an easy target for those who wanted to create a simplistic narrative of Darwin's historical advance. Yet, (1) there is no straightforward contradiction between Owen's representation and an evolutionary or phylogenetic understanding of taxonomy — species need not be posited as eternal, but only as the historical and contingent variations of these archetypes; (2) Both Owen and Darwin, both archetypal and evolutionary thought, require the same distinction between homology and analogy — neither is more empirical or essentialist than the other, but rather history can only be made to appear once essential and inessential have been filtered out.

This typological thought may be everything certain evolutionists claim to despise today — "essentialist," "idealist," "Platonist" — yet it represents a necessity on which their work is just as dependent. "True" relatedness does not simply appear on the surface of reality, to be skimmed off by an empirical gaze, but is dependent on processes that are no more simply material than ideal, essential than historical, given that essence, history, matter, and idea are derivatives or products of the "same" processes. The best indication of this in the contemporary biological field is the reemergence of certain forms of typological or structuralist thought. This apparent renaissance is tied in part to displacements of evolutionary theory that have been brought about by the study of development. Evolutionary biologist Günter Wagner has suggested a reformulation of the concept of homology, not simply as the representation of relations of descent, but of derivation from a common set of developmental possibilities ("Biological Homology"; "Developmental Genetics"; *Homology*). There are important structural relationships that cannot be captured within a historical categorization, including relationships among structures within a single organism — Wagner's theory thus rescues a concept that was important to Owen, that of *serial* or *iterative* homology ("Biological Homology" 52). Wagner argues as well that only his theory can capture the homology of sexual difference — the relatedness of structures such as the penis and clitoris ("Biological Homology" 62). Wagner refers not to common historical descent, but to shared "developmental constraint" to categorize these relationships; he designates this reconceptualization, perhaps superciliously, as "biological homology." Because, in Wagner's phrasing, certain structures are "built anew" each generation, it makes little sense to ask, for instance, which digits of a five-toed animal "descend from" which digits of a two-toed animal, or vice versa. Rather, by studying developmental systems, one can come to understand how a single network of possibilities can lead to the diverse outcomes of, for example, limb development. That is, the relation of homology is not a straightforward result of inheritance but categorizes those structures that develop out of the same essential possibilities. This research project (EvoDevo) has led to the astonishing discovery of largely conserved gene regulatory networks guiding the development of organisms as diverse as fruit flies and human beings. These Hox or homeobox genes are responsible for the development of morphology, and thus represent, in a certain sense, the rediscovery of archetypal body plans.[11]

My goal is not to pretend to privilege a "Platonic," "essentialist," "idealist" representation of life over any other, but to suggest that the aporetic relations of history and essence do not form a simple opposition — there is no historico-evolutionary discourse of immanent becoming that could be set against some adherence to essences. History implies the question of essence, which nonetheless is only posed historically. It takes only a single step, beyond the necessarily simplifying history I have recharted here, to recognize that these gene regulatory networks thought to encode the archetypal or typological possibilities of the *Bauplan* must have a history — they certainly did not fall from heaven fully formed, and their very universality can be conceptualized as a result of the common descent of the living. Thus, "historical" and "biological" homology are not *opposed*; they imply each other while corrupting in principle the possibility of a simple and uncorrupted principle or origin.

Despite the massive transformations separating these epochs and structures of thought, they all face a similar aporia. In every case, the rational, historical, or structural relation must be distinguished from an imitation that could be called *analogy*. This relatedness, both a failing and an enabling condition of anything like a system, is perhaps the ultimate mark of their common descent. In the case of analogy, we are again dealing with something that resembles or imitates vertical descent. Yet, it is not exactly a case of horizontal transfer — in fact, analogy only imitates horizontality as well. This analogy, its ersatz filiation or infection, is a sort of virus of the mind. It is in a sense an artifact of the investigation, only there because of something like a misrecognition of the genealogist. Yet, this "misrecognition" is more fundamental than the origin itself. It is the origin of the origin. Without passing through this naïve moment, from which an origin can be sought or deduced on the basis of resemblances among its effects, no origin would ever be discovered or recovered. This ana-originary or anoriginary origin makes all proper and improper filiations both possible and impossible by making them descend from what is in principle heterogeneous to essence.

One way of recognizing the aporia of homology and analogy is to find ourselves caught in the following circle: we restore the linearity of history (or descent from the essence) by distinguishing the homologous from the analogous, identifying ancestry by its effects, yet there is no way to separate the wheat of homology from the chaff of analogy without *knowing that history already*. In the words of two important biological systematists of the second half of the twentieth-century, "[Homology] is problematic in that true (taxic) homologies cannot be distinguished from false ones (homoplasies) without some assumption of hierarchy: homologies are more often deduced from trees than trees are from homologies" (Doolittle and Bapteste 2045). Only a hypothesized or naturrated history creates the distinction that inaugurates it. Which is not to say that we start from nothing, creating history each time as if from a bank slate — rather, this task or demand to sort being from seeming is itself inherited, from a source we cannot master because "we" too are one of its hand-me-downs. While there are certainly more and less rigorous ways to reinvent this history, none of them brings us in principle outside this circle. For instance, while we can search the fossil record or phylogenomics for a propaedeutic to this history, there too homology must be distinguished from analogy, and one begins only with resemblances or a-filiations of uncertain value.

There is nothing empirical, nothing gleaned from the surface of the given, that could distinguish homology from analogy. Rather, the distinction descends from the stories we tell.

Still, this seemingly insurmountable principle can never be more than an analogy of origin, a resemblance or substitution caught in the torsion of a fold. Analogy will only ever appear to us as analogy to the extent that something else is preconceived as proper identity, with everything that implies about true descent from the essence. Even if it can be shown that we never recover that essence or origin without passing through what we call analogy, the schematization and intuition of essence-effects, "analogy" could not be understood as such unless it had already been contrasted with a pre-definition of a real unintuited cause. We can even, by means of something like a higher analogy, attempt to ground our intuitive analogies in this "real" causality, by seeking the chemical and physical causes of our perceptions and thoughts, and even the narratives of evolutionary descent that could explain why we observe certain things as related, often despite what can be confirmed by mechanical analysis.[12] Our narratives or naturration of life's history are constructed by something like an analogy, a primary displacement that is unidentifiable and powerless without those very narratives — metaphysics and phenomenology turn in this circle without being able to measure or schematize, to name that other that must set their turning in motion.

Trees, Webs, Networks, Rhizomes

Facing the undecidability of vertical and horizontal, homology and analogy, is anything but the idle pastime of recondite theorists. Precisely these questions are at work transforming the models of life employed by practicing biologists, who receive these deconstruction-effects with varying degrees of resistance or hospitality, naïveté or cunning. For instance, these effects appear in the work of Eugene Koonin, whose definition of the virus was introduced above, and whose research broadly explores comparative genomics to probe the most fundamental questions of the origin and evolution of life. His consideration of the Tree of Life brings him to precisely this impasse or undecidability: "Herein is the fundamental problem that literally reaches philosophical heights: to meaningfully speak of HGT [Horizontal Gene Transfer], one *must* define the 'main,' vertical direction of evolution. However, if organisms exchange genes at high rates — in the extreme, freely and uniformly — the concept of vertical evolution makes no sense, nor does the orthogonal concept of HGT" (Koonin, *Logic* 124; original emphasis). It follows from horizontality that the vertical *never was*, for all the reasons I have suggested, yet there is no horizontality to speak of without a tree-structure in place behind it. Koonin's invocation of philosophy, to resolve this aporia (or absolve him of his responsibility for it), is symptomatic of undecidability. He tries to repress the most disruptive consequences of a deconstructive specter by pretending that it properly belongs outside his own field (in the empyrean of "philosophical heights"). In other words, a scientist who recognizes that both the rational and the natural grounding of their own discourse are impossible, that they *must* study their own fictional reconstructions because nothing exists without this supplement, may always pretend to negotiate with this intrusive thought by acting as if it belonged to another discourse (and thus, as if they could continue their work unimpeded by it). In a certain sense it does — it certainly does not belong "properly" to biological or scientific discourse given that its impossible remanence cannot simply be exposed or exhausted there. However, it does not properly belong anywhere else, either. This ex-heritance or a-filiation, which makes every descent descend from what it is not and does not belong to, necessarily remains over in any discourse, whatever attempts one makes to draw or suppress its consequences. It will always be *as if* another discourse, taking place somewhere else, for some other purpose and in some other language, could speak directly of what troubles us here only in the form of a certain indirection and counterpurposiveness. A virality. Still, the double lure of this apparent alterity, or rather this in-apparent haunting, should not be thus cast aside, for reasons that I will somewhat artificially enumerate:

(1) There is no other discourse, language, or field to which it properly belongs. One will not find, within philosophy or any other domain, the answers or even the resources to address the irresolvable questions opened by the relative non-relation of a-filiation. My discourse here, for instance, is just as "philosophical" as it is "scientific," and if it is not straightforwardly recognizable as either it is only because it dwells on these haunting remainders that cannot be exhaustively internalized or domesticated by scientifico-philosophical thought precisely because they have never simply been external to it. What exceeds science and philosophy, in this and every discourse that speaks their names (whether it invokes these fields to belong to them or pretend to surpass them), is nothing purely external to them. The questions that have started us down this path or into this looking glass or mise en abyme — what distinguishes "vertical" from "horizontal" relations and thus essence from accident, true knowledge from mere appearance, subjective intuition from objective cause, how to ground the apparent contingencies of history in law — these are questions without which nothing worthy of the name philosophy or science could even take its first step. Yet, each of these questions, which are not simply separable but shine through each other like facets of a crystal, opens on something that leads *beyond itself of itself* — the strictest order of rational inquiry, whose logical and epistemological method is also in question here, cannot help but lead to its own deconstruction, to a vantage from which what seemed to be our progress and the tools clearing our path and providing our landmarks are rendered delirious. It is reason "as such," philosophy "as such," science "as such" that corrupts its own principle, and thus the *as such* as such, the principle of every principle, that affirms its authority by dissipating it, relaying to an other who remains to come.

(2) As cause or consequence, if these questions or problems, these aporias, do not belong somewhere else, it is because they suffuse the entire "interiority" of any field striving for rational self-enclosure, of anything that would count as a field. We are measuring or undergoing the ordeal of a sort of weak force, which cannot be identified with the horizontal or the analogical any more than their supposed others, but is the fictiveness or virtuality of the a-filiation that makes these relations both possible and impossible — this weak force, whose effects appear to us as a sort of glitch or ill-cut puzzle piece, do not leave a single unit of the supposed interiority of these fields untouched or intact. Every instance of inheritance of and by the living has its interpretability opened to irresolvable questions by a-filiation. We are not facing a question of the origin of life, if that means one whose only effects took place almost four billion years ago, but a specter haunting every possible and impossible conclusion made by or about the living — every act by which the living hope to accomplish their reproduction, and every conclusion in every text and textbook on evolutionary science are held in suspense, speak with a borrowed voice that echoes like the illusory reverie of a narcissism — *you are my evolution . . . lution . . . lution . . . lution . . .*

These problems or aporias do not belong outside science, but they are not simply internal to it either. There are certain signs, to which I will turn in a moment, that Koonin imagines he can defer these problems or their consequences for treatment in another locale, or even resolve them within his own corpus. Still, he recognizes that they expose a sort of opening or wound to infiltration by parasites. In response to scientists who dismiss the evidence of Horizontal Gene Transfer to cling to the most traditional representations of a Tree of Life, Koonin can find no way to describe the situation other than acknowledging, to an extent, that something extra-scientific is intruding upon apparently scientific debates: "With the accumulation of comparative genomic data, the anti-HGT stance is quickly becoming more of a psychological oddity than a defendable scientific position" (*Logic* 148). Koonin is suggesting that, because observable facts challenge the Tree of Life model, those researchers who cling to it are beyond the reach of scientific argument, in a zone that would require someone like a psychoanalyst to search for the navel of their dream of arborescence. In a sense this is correct, in that no observation can ultimately prove this model (everything that will count as observable rather descends from it), but Koonin is too quick to assume that this attachment to auto-disproving interpretations is an individual "oddity" or pathology, rather than a constitutive necessity of science and every other form of thought. Koonin imagines that there are diagrams or models adequate to a "horizontal" world, even though he has already correctly deduced that *horizontality undermines its own possibility.* That is, a model of horizontal relations such as a web or network is no more and no less fictional or virtual than a tree-model. Thus, nothing the scientist or anyone else can put forward as observation or interpretation will ever be sufficient to determine what sort of model should represent the living — they will always be thrust beyond their own field, toward the psychological and anything else, to try to justify what, according to the authority of their own method, is unjustifiable. The question of tree or web is not merely one of *what is*, but of *what is desired* — why some of us hope against hope that we stand upon its branches or within its threads. Every supposedly "non-scientific" consideration, such as the "philosophical heights" or "psychological oddities" mentioned by Koonin, shows up within the fold of "science" "itself," once we have recognized that the scientific is grounded in everything it is not — for example, in what a given investigator thinks a natural family *should be*.

Koonin himself is not immune to this desire. We just read his acknowledgment that horizontal exchange is no more coherent a concept than vertical inheritance — like a sort of Cretan its very possibility renders itself illegitimate and impossible. Nonetheless, Koonin immediately follows his description of this horizontal paradox by writing: "Hence, a web (network) representation of the evolution of prokaryotes seems to be a logical necessity" (*Logic* 124). How can any sense of "logical necessity" be salvaged once the consequence of a premise is its very undoing as a premise? The web or network of life is just as "illogical" as a tree, once we have admitted that there can be no stable divergent hierarchy (tree) among the living, and thus no Archimedean standing point or ground from which to judge and measure the horizontality of its transfers. In other words, every web is the cross section of an implicit tree. The most programmatic statements of biologists who attempt to reconstruct the "net of life" make its dependence on tree-structures apparent — for example: "Here, we present a first attempt to reconstruct such an evolutionary network, which we term the 'net of life.' We use available tree reconstruction methods to infer vertical inheritance, and use an ancestral state inference algorithm to map HGT events on the tree" (Kunin et al.). To construct a net or network, step one is to construct a tree. Just as before, when the investigator was forced to separate verticality from horizontality and homology from analogy in order to identify the branches of their tree, it is necessary in order even to begin to figure life as a network that one decide what will count as a node and what will count as its edges (its connections to other nodes). Every possible arrangement and rearrangement of these relations can be explored, but no figuration at all is possible without granting some degree of stability and identity to these nodes and their relations. The web, network, or rhizome is just as fictional and remains as deconstructible as any tree — *because they are trees*.[13] They are not, and never have been separate or separable representations of life. There is no possible figuration, however systematic or impressionistic, that does not rely on this fictional stabilization; life, if there is any, descends from the unfigurable possibility and impossibility of all figuration.

Readers of Deleuze and Guattari will have recognized that Koonin's concerns overlap with their anti-oedipal or rhizomatic project in several particulars (Koonin himself cites them). Deleuze and Guattari introduce the rhizome precisely to pursue an order of horizontal relations that could displace the trees with which they famously declared their boredom. This rhizome has even been taken up in the work of contemporary evolutionary theorists, some of whom have tried to attach it to distinct formulae for the filiations of life (Raoult; Koonin, *Logic* 422; Merhej and Raoult; Georgiades and Raoult).[14] While a reading of Deleuze and Guattari's entire corpus would be necessary to position their work relative to what I am articulating here, I believe such a reading could show that their attachment to productivity, to their own philosophy as the "creation of concepts," leads them to reify tree structures, the better to celebrate the rhizome as an escape from them — an oppositional relation, rather than a haunting of undecidable alterity.[15] Such oppositionality can be felt everywhere that they "invent" binaries in order to critique binarism, and suggests effects that are as questionable in scientific or ontological terms as they are for their ethico-political tendencies. This implicit ontologization, which reifies borders in order to celebrate their crossing, is presupposed everywhere that some form of becoming, of symbiosis, relationality, plasticity, process, or network, operates as a privileged term

Nothing like horizontal or vertical relations take shape without some sort of fabulation or naturration. Still, this narration or textuality is not freely imposed on the formless but is itself the legacy and reconstruction of an inheritance, including but not limited to our inheritance of names.[16] Even if an adequate narrative of life is impossible, there is no outside of this narrativity — we arrive at ourselves as a link in its chain. In other words, it is not because life is a supralogical power that we are condemned to replace it with substitutive stories, but because we will never surmount the necessity of beginning from "life," from what is only a given name in a particular language, an inheritance in which we are both the recipient and the gift. There is no evolution without the transmission and translation or traduction of names: there are no kingdoms, domains, species or organisms, and no ruptures or transitions between them, without this nomenclature or graphical supplement. Nothing on the immediate or intuitive surface of experience, as if there were such a thing, tells us if something is the same or different (a reproduction, a speciation, etc.). The lesson of the horizontal and vertical, the homologous and analogous, is that the same is not necessarily the same nor the different different — one must return from the ephemerality of this surface to something that undergirds it like an essence, a narrative of becoming that is always already underway and that is capable of a-filiating the most similar and dissimilar things.

Alongside those "psychological oddities" that Koonin sensed lurking behind his colleagues' arborescent constructions, it may always appear as though certain effects stem from unexpected depths of our inheritances. We have recognized that the "Tree of Life" tries to unify the logical and natural relations of the living. It is not irrelevant to point out that it is often written with initial majuscules in the text of biologists, as if it were a proper name — as if there were only one, or at least as if a unique one were named every time.[17] Even if there are as many different proposed trees as there are scientists speaking its name and seeking its root, in principle there is only one, the Tree of Life is the name of the unity of life, its belonging to a single natural family. There is more than one reason to suspect that this search for lost unity has its roots in the biblical story of Genesis, where God planted in the Garden of Eden a Tree of Life from which Adam and Eve were barred after eating the fruit of the Tree of Knowledge of Good and Evil — *lignum scientiae*.[18] Along with the restoration of life to a single family without rupture or discontinuity, without death, the Tree of Life is the messianic hope of healing this apparent divide between science and life.

Those who hope to return to a nest or cradle among its branches find it is the Tree of No Ledge.

Is the tree a natural or logical figure? Could we understand the relationship of premise and conclusion without learning the force of generativity from nature? But could that generativity ever appear to itself without first being brought under a logical yoke? The circle formed by these questions is a ring in the trunk of the Tree of Life. It is impossible to say whether this is a natural figure, displaced to represent logical relations, or whether something properly logical is making nature accessible. No matter how self-evident it may seem that a tree is first and foremost that living something in nature, it is impossible to grasp the idea of its or any species-being without starting out from an at least implicit branching logical form. As if something like impropriety "itself" were the condition of possibility of the proper. Still, our task is not to try to assign priority, but to read logic and nature as the traces of a-filiation. For instance, the risk that must be avoided is that of thinking that if "tree" and even "Tree of Life" is something *logical*, this means that something truly and purely *natural* will be found somewhere else, untouched by the supposedly benighted investigators who mistook its true nature. This is implicit in every discourse that imagines it can distinguish natural and logical form, to determine if a logical tree is adequate to the form of nature, or an artificial imposition upon it: "Questions about the structure of the TOL [Tree of Life] are, after all, secondary to questions about whether such a branching pattern actually corresponds to anything in nature (rather than being imposed on nature by the habits of systematists), and if so, whether a branching evolutionary process is its underlying cause" (Doolittle and Bapteste 2043). This very distinction, between what is "actually in nature" and what is "imposed" upon it by rational systematicity, cannot possibly pass judgment on the tree form *because it is itself a branching dichotomy.* Every attempt to distinguish tree from non-tree can only create another branching opposition, another tree. In this case, while pretending to have at hand "nature," such that it could be compared to a treelike logical or rational form, these biologists have unwittingly made nature "itself" the descendent or branch of a tree, one supposed to root the opposition of "nature" and "logic." It is no longer a question of attempting to distinguish what is "actually in nature" from what has been "imposed" on it, once we recognize that there can be no nature prior to inscription, an arbitrariness (arbor-trariness) that appears as the abstractability of logical form.[19] A nature that never was must be selected just for us to understand artifice as its deviation. Nothing that moves in this circle has the givenness or pure productivity that a traditional logic might associate with the generativity of nature or the concepts of the understanding. If it is naïve to expect that nature will simply generate or give itself in the form of a network or web, it is just as naïve to think that a scientist or anyone else among the living will produce or create it from whole cloth. It is received as an inheritance, which means it is exposed to every undecidability and aporia of a-filiation.

If the Tree of Life is the most artificial thing, dependent on seemingly contingent details of this history and its languages and cultures, there is nonetheless nothing more natural than this artificiality. Nothing is more universal, innate, congenital, and original (thus natural) than this dependence on the contingent, particular, conventional, symbolic, and so on. More fundamentally, even the historicism that determines these factors as contingent by placing them in their context depends on something it cannot account for, and which undermines its resources even as it grants them — any history presupposes naturration. Even if everything that could ever appear as a tree, and every word or symbol that could ever stand for it, always manifests with the contingency of its particularity, there is nonetheless something like an inner alterity, this haunting of the proper by what it is not, that seems more necessary than nature "itself." This something, which is neither "the" tree, nor something other than it, is beyond truth and falsehood — it is not simply false, given that there is no truth without it, though its interminable deconstructibility rejects the yoke of any truth, carrying the germ that infects every truth with falsehood — and, like humanity before the fall, it is beyond good and evil — it would be impossible or somewhat fatuous to celebrate or lament this something, to say we were glad for it or tired of it, given that there is nothing without it, whether the worst or the best. If there would be nothing at all without it, perhaps that is something.

It could be shown, as well, that death could not afflict the living without the supplement of some tree. Nothing could end unless it were first possible for it to inherit its identity in a process that makes it relatively conventional or symbolic. The punishments befalling Adam and Eve for their transgression or their knowledge, the toil and labor that maintains life only from its own artifice, the pains of reproduction and the *peine de mort,* continue to befall life as the *malum* of some tree. Everything that is expressed in a mytho-poetic or narrative fashion in the story of the Tree of Life (which has many versions beyond the Abrahamic traditions), can be rediscovered in a relatively scientific or philosophical avatar of this tree.[20] Nor is it accidental that the form biology gives to descent among the living suggests congruent figures such as the "mitochondrial Eve" or the "Y-chromosome Adam" (Oikkonen).

I have focused on Koonin's work in part because he recognizes, to a degree, certain fundamental disjunctions that haunt every biological project. Still, he often passes very quickly from the destabilizing potential of an aporia to the acceptance of a stabilizing origin and narrative. We saw how quickly he passed from the auto-hetero-deconstruction of horizontality to the "logical necessity" of the web or network, which is precisely a horizontal structure. His historicism ultimately restores a treelike aspect as well (as, in a sense, it must), by taking the form of a "Forest of Life" composed of "Nearly Universal Trees" (NUTs) (Koonin, *Logic* 153–56). What appears as a certain desire to salvage this narratability may be in its own way a desire to inherit and inherit well. It is not without relation to the inheritance of names, and in this case Koonin's forename: Eu-gene. The well born, the good gene. The entirety of his work in genetics might be a kind of birthright, pursued with the compulsion of an instinct or a curse.

Loss as A-filiation: Gone and Epigone

We are not dealing, then, with separate or separable substances or even processes, some of which can reliably be identified as vertical and others as horizontal. It is not the case that within a chain of vertical transmissions occurring as history itself, horizontal transfers occur that erase the record of that history. It is only on the basis of an implicit or assumed history or nature that the horizontal can appear as horizontal, and without the exclusion or filtration of the horizontal, the analogous, the parasitic, and so on, there would be no history. The "vertical" and "horizontal" are stabilizations of an a-filiation by which the living *give what they do not have*.[21] Similarly, analogy cannot be dismissed as a mere artifact of investigation, obscuring the true relations of homology. Every homology begins as the observation of an analogy, or the restoration of links in the chain of an apparent dis-analogy, the restoration of *family resemblance*. And yet, despite corrupting homology in principle, making it descend from its other, there could be no analogy that does not presuppose a preexisting system of homological relationships. The very meaning of the term is something like false homology — if it nonetheless holds the *truth* of homology in suspense, we are without a name for what is neither homology nor analogy yet allows both to stand in its borrowed light. How can we describe this something that brings forth everything that is, every resource we could borrow to explain it, but brings forth only illegitimate offspring, an inheritance without resemblance or relation, a contingency without contiguity, a contagion without germ?

There is yet another pattern that complicates the histories and desires of geneticists. If one hopes to reestablish the true lineage of the living, one must grapple not only with the possibility that elements may be duplicated, mutated, exchanged, and imitated, but that they may be *lost* as well. This non-self-identity is constitutive of every reconstitution of history or essence — its translations are not at all specific to life, unless one defines life as Walter Benjamin did, as "everything that has a history of its own, and is not merely the setting for history" (255). Everything inheritable can be lost, including one's genes. Just as vertical and horizontal, homology and analogy, were not simple oppositions or alternatives but undecidables, loss cannot be placed against inheritance or continuity as a distinct possibility. One cannot reduce its destabilization to the seemingly decidable question: Is X unrelated to Y, or did X merely lose a feature that Y retained? Again, the temptation would be to imagine a historical process throughout which traits and genes either repeated or disappeared, the latter merely obscuring the historian's or paleontologist's attempts to reconstruct that past that nonetheless took place. Even if one were to bear witness, in the present, to an act of reproduction, and to leave a memory or recording of its exact process, such that its before and after could be itemized without error (clearly, we are indulging in science fiction), even if such a memory were possible, it would not be possible to simply point to an element that had disappeared and say, there it is, *loss*. Every disappearance may be the loss of some part of the essence, but it may just as well be the restoration of essence, the expulsion of some "horizontal" accretion, some virus or parasite. Again, like horizontality or analogy, loss only appears as loss on the basis of an essentializing history that does not exist until loss has been distinguished from restoration.

Once again, what appears to be a mere historical contingency becomes a condition of possibility of history "itself," while also corrupting it in principle. What would a legacy or inheritance be that could not be lost? Something that gave itself as a sort of law or program, a possibility present from the beginning and impervious to change, without the possibilities of horizontality, analogy, and oblivion, would never really be given to or from anything. Without this difference, by which it risks itself in the other, a-filiation would never have its chance. But — and this follows with the force of law even if it corrupts the principle of every law — as soon as this loss is possible, that is, always already, it will never be possible to recover an origin, an original cause, progenitor, or patriarch, that is not haunted by this difference. Even the absence of a mark is a potential sign of its legation. Neither descent nor infection, neither being nor seeming, neither presence nor absence, and yet impossibly giving all of them their chance; there will never be a name for what is not even nothing, given that this too is only a borrowed resource. There is no simple source at the origin — what appears as the simple is already a parasite of this nothing, achieves simplicity in light of its différance. It is no longer a horizon, an ana*logy*, or even a nonbeing, but its very unnameability leaves a graphic trace that can be *read*.

To be, to be within this circle or fold, is to inherit.[22] What kind of inheritance arrives from beyond anything that could ever appear as itself in the present, and cannot be relayed to the future despite forming the very fabric of that future "itself?" Given that any term will do, so long as we see through it, I will nickname what keeps us turning in circles — circles that never quite return to themselves or to what might thereby prove itself a starting point — let us speak of the *gone.* Beyond its deceptive familiarity to an English reader, this tetragrammaton transliterates the ancient Greek *gonē*, which is seemingly as far from absence and loss as possible. *Gonē* names generativity and its products: it can refer to a child, as well as the seed, organs, or act of childbirth from which a child springs, and the generation, race, stock, family, or parentage to which it belongs. Our closest equivalent might be *generation*, as it can refer both to an act and its product, though we are without a single term that gathers the same semantic range. Our word *gonad* is a descendant of *gonē*, and *genital* is its cognate, both from a root referring to any manner of becoming, of genesis. The gone can name all generation and filiation, but only because it stands for the nonorigin that gives them, impossibly, their chance.

The gone cannot communicate or filiate either vertically or horizontally — even if every "natural" reproduction bears its traces, it remains over as the very disjuncture that exposes birthright and dissemination to their risk and opportunity. Something older and younger than every ancestor, because it has never simply existed, places not only every generation's status as generative in question, but the whole of "nature," which risks being exceeded at every moment. If the living can be represented as a single inescapable family, that is only because the gone allows for a reclaiming by opening the abyss or rupture that exposes "life" at every moment to what it has never and could never be of itself or of its own power. The gone engenders nothing, has no power and no fertility, but only because it keeps nothing for itself, allows no gift, sacrifice, or praise to reach it, exhausting and overdrawing or overflowing itself at every moment merely by bequeathing its silence. It even has no form or mark by which it could be recognized and said to repeat, to maintain or reproduce itself, but hides in every form and name as the deferral that leaves a space for reading. For all these reasons, the gone cannot assimilate itself as a rude or parasitic invasive species, an alter-lineage that infiltrates or infects the proper horizontally. It certainly does not befall a preexisting tree or organism, nor does it divert a purposive self-reproduction toward a reproduction of the other (which would be, properly, impossible, given that it has no purely repeatable mark). The gone was always there already, not in an enduring omnipresence but as an unassimilable alterity that fictionalizes everything like a tree or true descent before it even begins — it leaves the true, the proper, or the vertical haunted by its inability even to infect itself.

The gone certainly will not be tied to any history of homology, but it cannot be schematized or intuited as an analogy either. It has no form, and thus no resemblance to any and everything that nonetheless evokes it or evokes its unimaginability. Nonetheless, even if every science and philosophy that shuttles between analogy and essence has nothing to say about the gone, everything that can be said ventriloquizes it.

It might seem that gene loss, or loss in general, grants access to its true nature, but absence and nonbeing are inimical to the gone. Only elements defined by the system, definite or defined absences potentially contained in the combinatoric of its program, can be lost. Thus, every absence seems to have been already present, and even retains a kind of claim over a lineage. Nothing is ever lost or gained by a system (even if this means that no system has ever existed) — nothing happens in it or to it. The gone, without being present here, is nowhere else — its alterity is not a possible presence but the proximity of the impossible, threatening or promising to overturn every seemingly stable system.

Nor, in the end, will it have had a name. Gone is not a word, which is the only reason anything at all, here or ever, can be written or read.

Which does not mean that "we" are the appointed few who decipher or gather its traces — we who are those very traces. We, the epigone.

Chapter 2

The EVEs of the Genome

The previous chapter explored the inherent instability of classifications that separate "vertical" genealogical relationships from the "horizontal" transfers that imitate them by traversing the lines of a family tree. This chapter observes that same instability within attempts to conceptualize the virus as either an agent of heredity or of contagion.

Much of this chapter focuses on viruses that infect bacteria, known as bacteriophages. *To understand the full extent of their subversion of self-identity and straightforward (oppositional) alterity, I examine what are sometimes called the two "life cycles" of such viruses, the* lytic *and* lysogenic. *The more familiar lytic cycle involves a virus injecting its genetic material into a cell, using the cell's machinery for copying genes to rapidly reproduce, and then bursting (lysing) the cell to distribute virus particles (called* virions*) to infect more cells. Such an infection spreads rapidly and "horizontally."*

The lysogenic cycle involves a virus leaving a copy of its genetic material in a latent form in the cell, so that the virus spreads in a "vertical" fashion, copied into the cell's descendants like any other gene. These lurking viral genes are known as prophages *and can be integrated into the thread of DNA containing the bulk of a bacterium's genes, sometimes called the bacterial chromosome or* genophore, *or can form an independent loop of DNA sometimes called an* episome. *A distinction is often made between viral episomes and other DNA structures that float in a bacterial cell's cytoplasm, called* plasmids, *although this distinction too is tenuous.*

Even for multicellular organisms with a segregated germ line, meaning that genetic changes can only be passed on if they occur in the specific cell lineage with reproductive potential (sperm or eggs), it is possible for a virus to integrate its genes into a germ cell and spread in a "vertical" fashion, from parent to child. These hereditary viruses are sometimes called Endogenous Viral Elements, or EVEs. *They comprise significantly more of the human genome than those segments typically referred to as our "genes."*

***

It was argued in the previous chapter that classifications separating vertical relations of descent from horizontal relations of contagion were inherently unstable. It will follow that the virus is not simply identifiable or classifiable within any such system. It may nonetheless be possible to construct trees that make it the origin and progenitor of life "itself," or to figure it horizontally crossing and tangling the branches of a filiation it seems to corrupt from the outside and by chance. Still, it would be naïve, after demonstrating the deconstructibility of every possible family tree, to pretend that *the* virus was the agent or germ that just happens to sow disorder and confusion among the otherwise stable relations of the living, let alone the fixed root of an unshakeable tree. The movement of deconstruction could never come to rest in this way on a particular agent of disorder. To make of the virus that which infects the otherwise immune is necessarily to imagine it as something with an essence of its own, belonging to its own family structure or possessing its own family resemblance. Picturing the virus as an agent of horizontality or infection makes it more living than life "itself," which means it is just as deconstructible. That is to say, there is no *the* virus; it forms neither a natural family nor a coherent concept, has no essence and no identity, no species and no individuals, neither living nor dead. There are *viral effects* only because something that is neither life nor virus, neither vertical nor horizontal, makes every family tree both possible and impossible.

If the virus seems to be continually turning the Tree of Life on its side, rotating it and replanting it by its branches, that is because it is no longer possible to fix the viral in either a homologous or analogous relationship to vitality. It is just as possible to claim it as an origin and extension of life's self-accomplishment (we will see that it plays a role in birth, immunity, genetics, and speciation), and thus a component of its heredity, as it is to claim that it *merely* imitates life, infecting life's supposedly autonomous accomplishments with an alien force. This undecidability follows as soon as one acknowledges that the virus *isn't anything* — there is no *virus that would be one*. There are neither material structures one can identify it with nor stable predicates for its conceptual definition, but the seemingly self-same is continually transmuting from self to other in a fashion that prevents rigorous anticipation or delineation. If there is no family or concept of the virus (nor one of life, a subject to which we will return), it will not be possible either to identify it with or oppose it to the living as its power or counter-purpose. Life "itself" is a bad imitation, a resemblance or analogy of itself or of what it itself is not or should not be, and a "*mere*" resemblance because it does not share an origin with itself but comes to be by a supplanting and fictionalization of its "original" nature — a nature that ought precisely to be originality and true descent from the origin. The virus too is a bad imitation — such a *bad* imitation, in fact, that it even *imitates poorly* by being too much like what it would have to not be, itself or life itself, if we would hope to salvage some non-imitative or true life.

The virus has no simple origin. We can already anticipate this complication given that, even or especially in a discourse that places in question its inheritance of names (how else to know if we understand anything or anything true when we use this "word"), one still must guide one's thinking with the very terms thus held in suspense. It would not be possible to say when the virus was first discovered or defined, because it was only in hindsight, with the stabilization of the category, that discovery seemed to have already taken place. It would not be without reason to refer to this process of displacement, an origin taking place *nachträglich,* long after the fact, as virality — even if we do not know what that means, and as yet another sign that we are dealing with something that always precedes itself without ever simply arriving. It will not be possible, then, to say "from the beginning . . . ," to speak of a first virus or a before the virus — rather, wherever we look for a beginning, we will already find virality in play; we will find ourselves placed in question. Some form of the question of whether the virus belonged to life or befell it from the outside, and thus whether there was a "virus" to speak of, could be said to have always been there. For instance, in 1915 Frederick Twort published a paper detailing a phenomenon he had observed in bacterial cultures, and which would only much later be viewed by consensus as viral (we will return to this history). What we now call viruses were then too small to be seen by microscopes, but Twort could observe transparent spots appearing in his bacterial cultures, as if an infection was spreading among them. While he entertained the hypothesis of an "ultra-microscopic virus" (virus being then a generic word for a pathogen), he preferred a different explanation, in part because of an observation that was difficult for him to square with the notion of infection. Cultures that were apparently pure of contagion and grew normally for several generations would eventually, seemingly spontaneously, develop these epidemics:

> There is this, however, against the idea of a separate form of life: if the white micrococcus is repeatedly plated out and a pure culture obtained, this may give a good white growth for months when subcultured at intervals on fresh tubes; eventually, however, most pure strains show a transparent spot, and from this the transparent material can be obtained once again. Of course, it may be that the micrococcus was never quite free from the transparent portion, or this may have passed through the cotton-wool plug and contaminated the micrococcus, *but it seems much more probable that the material was produced by the micrococcus.* Incidentally, *this apparent spontaneous production of a self-destroying material* which when started increases in quantity might be of interest in connexion with cancers. (Twort 1242; emphasis added)

Viral infection was indistinguishable, already and in principle, from the "apparent spontaneous production of a self-destroying material," a sort of autoimmunity. As our knowledge of viruses has developed, it has become more difficult, rather than more straightforward, to distinguish them rigorously and in principle from their hosts.

To deny the natural or logical cohesion of the virus is not at all to pretend that what we call infection does not take place, nor to deny that within certain limits its causes can be identified, recognized, and extirpated; nor would I deny that there are systems of immunity worthy of protection, systems we would call living — all of this can and has taken place without there being any such rigorously definable thing as *the* virus, without us being able to identify its nature or distinguish it conceptually from that life it is supposed to threaten and subvert. A deconstruction of virality and vitality is as far as possible from casting everything that has gone by these names into a pit of indifference, or a night-in-which-all-cows-are-black, and is anything but a denial of the ethical differences that attach to everything so named. It is only out of the greatest possible respect for something like the *reality* suffusing this field, and even a respect for the most rigorous possible philosophy and science that would approach it — it is "realism" itself that places these terms in question or causes them to tremble, leaving us without sure footing as to either the subject or the object of their investigation.[1] If there is something irreducibly unfamiliar, or even uncanny, about this "reality," it is doubtless that it is no longer simply *opposed* to a fictionality or virtuality that grants its only possible standing while also promising that this name too, the "real," and the security we have invested it with, must be placed in question. That something without prior or originary existence only borrows a name, a concept, a form, and a function; that it takes up only temporary abodes and yet for that very reason can be excluded nowhere, that it smirks silently behind the face of our own self-recognition; this is virality "itself," a movement of generalized infectivity that unsettles any resting point in a proper, immune, and self-identical corpus.

While exploring this undecidability of self and other or virus and gene, I hope to draw a firm distinction between what is underway in this text here, which might be called the deconstruction of the virus, and pseudoscientific writing by conspiracy theorists who have sought to capitalize on the confusion and fear created by the COVID-19 pandemic. Deconstruction is not at all, in fact is as far as possible, from the denial of the possibility of contagion that one finds in these discourses, nor does it prevent us from calling out opportunistic misinformation, or from privileging other representations in the name of their truth or ethical significance.[2] If I have suggested that every distribution of subjects and forces into self and other or immune and infected has a conditional or contextual dependence, and thus can be rearranged or rewritten, this is not at all to pretend that everything fades into indifference or that all such partitions are equivalent. First of all, this dependence that makes every "autonomy" an effect of context is insurmountable precisely because there is no outside of conditional arrangements and the charged investments they carry for "us" — one never steps from these conditionals into a circumstanceless zone of absolute indifference, but only into another context and another inheritance of investments (for instance, from an investment in personal health to an investment in communal or global health, natural or ecological health, and so on). The only *unconditioned* is this inescapable conditionality, which holds open the promise of deconstruction — not as something a critic does in the form of a text (in the traditional sense) — but as an event befalling anything we could count as a "thing itself." Nor is this conditionality a negative check or hindrance restraining something like the machinament of nature from outside, but is the only possible opening and chance for anything at all. This deconstruction carries, channels, or diverts something like desire and ethical law, or may even place in question what we believe we have known under those names, but it can only do so by raising the specter of another, inexhaustible demand. I would even say, in a dictum that redoubles on itself: deconstruction *should* always be practiced in such a way that the negotiated certainties of moralism or good conscience, which bargain with the constraints of existing injustice to seek self-satisfaction in what our "freedom" makes *possible*, tremble in the face of a responsibility that is all the more urgent for being *impossible*.

Second, following again directly from the necessity of context (one could say the naturality of what nonetheless manifests as symbolico-conventional culture and contract), whatever appears as a contextual element is never simply outside the system it seems to make possible and thus is destabilized along with it as conditions of possibility are made relatively explicit. One can certainly displace scientific structures of thought and practice by re-centering what seem to be their purely contextual conditions — for instance, language; culture; the subject position of the scientist; geopolitical borders; socio-political institutions and their perceived trustworthiness; a scientific establishment of journals, societies, universities, and research institutions; military and political investments; media framings and popularizations. Any science worthy of the name, any research striving for true knowledge of cause and essence, will necessarily find itself implicated in such displacements in a specular and vertiginous fashion. Among the structures held in suspense or volatilized by the deconstruction of life's vertical and horizontal relations is *everything that takes the form of an inheritance*; not only in "nature," which may indeed be less natural than we thought, but every one of these socio-politico-linguistico-cultural institutions that thus cannot *merely condition* the study of "nature" but are conditioned in turn by every stabilization or displacement of its basic structures. Thus, these institutions cannot simply pervert or distort a "view" of nature, but are outside it within it, implicated in that study without being controllable by it, children that make their parents possible. They are swept away in the same movement of deconstruction that they seem to initiate. A certain mirroring and anxiety over borders will always follow from this and never ultimately be settled — every scientist will be able to claim that the psycho-socio-political conditions of possibility of their research belong to their science, as simply a later stage in the evolution of matter or life, while in turn every sociology of science and a nonfinite set of possible or actual disciplines seeming to arrive from outside science "itself," will nonetheless, even as they turn everything on its head, do so precisely by carrying toward a certain limit the project of science in a fashion that need not be considered *unscientific*. And thus, are not at all immune or external to the displacements they seem to bring about of their own power. Rather than casting all commitments to the truth and the ethico-affective into an abyss, deconstruction destabilizes the terrain whose fixity might seem to offer grounds for disinterested and objective research — all science, as well as any text that speaks of science as if from the outside, takes place within a scene of family relations and inheritance that is as affectively charged as any other.

All this is to say that a deconstruction of life, the virus, science, or anything else would never, and couldn't if it tried, wallow in a nihilism of ethical indifference. In fact, there can only be "indifference" from the point of view of the inscription of differences it ignores or obscures, and it will be precisely because that inscription remains legible and deconstructible that indifference can be recognized as such and thus denounced, overcome, complicated, and so on. If the nature of the "virus" is here placed in question, and the extent to which we know what that is and can rigorously distinguish it from life, that is not at all to deprive those who call themselves living of decision in the face of immunity and everything that threatens it — nor is it to claim in a relativistic fashion that every determination of what should count as health and what as health-risk is equivalent — but precisely to exacerbate the responsibilities that must be taken and not taken for granted in the face of what is never simply a programmatic application of rules or authority (that of the scientist or anyone else). Only by bad faith can one absolve themselves of the responsibility to decide, by pretending that any old representation of life and its others is just as good or at least that we lack the power to arbitrate, given that one is *necessarily* already acting and judging on at least implicit principles (even or especially when one does not admit this to oneself) that polarize this field and make this decision in ways that implicate not only one's "own" life or possibilities but those of virtually everyone else. It will always be necessary to decide, within a certain stabilizing and thus destabilizable or deconstructible context, what should count as life and death, my own and the other's, health and infection, reasonable precaution and irresponsibility, truth and misinformation or pseudoscience, and so on. Again, precisely because of the deconstructibility of every context that maintains these formations, a ceaseless vigilance is called for at every moment, one that is not opposed to being overtaken by the unthought other, but that suffers its own responsibility most acutely where the limits of its powers of decision or response are placed in question.

Nor should we be surprised if no unified expertise and authority, no established distribution of responsibilities and certainties, proves sufficient in the face of what we sometimes call "crisis." Everything I have said about contextuality entails that no one will ever settle or arbitrate the relations among, for instance, those who understand best and thus are authorized to study and make policy regarding the "natural" processes and risks determining the course of a pandemic and those who can predict, legislate, or ameliorate its "cultural" determinants. Among the many upheavals it has brought about, COVID-19 has demonstrated that the pathways followed or forged by a virus, by viral *frayage,* are by no means restricted to those thought of as natural or biological — everything takes place as if a virus could *read* us, could trace, exploit, and even erect and reinforce symbolic and culturo-conventional or geopolitical divisions within and among societies (R. Benjamin; Hatch; Thrasher). The borderlines among differences of race, gender, sexuality, class, age, disability status, immigration status, and nationality not only channel the virus but can also be drawn and maintained by it; this viral differentiation is not limited to those vulnerabilities a medical scientist would attribute to our differing immune systems, but is inextricable from the social and political determinants and effects of pandemic — for instance, working conditions, unemployment, the increased demands of uncompensated social reproduction, houselessness, and incarceration — and all the forms of exploitation that take place within and between societies, including the legacies of colonialism in the Global South, international intellectual property law, and its weaponization by corporations whose wealth and power exceeds that of many states, determining access to vaccines and treatment. It is no more a natural than a cultural phenomenon that these differences can both be modes of transmission, and be redrawn or displaced by a virus, such that symbolic events resembling speciation can take place in the fields of race, gender, class, and any form of *Geschlecht* as a result of viral contagion. We might displace the genetic understanding of certain of these categories by framing them not as determined by reproduction and innate inheritance but as effects of shared vulnerability to infection — in the generalized sense — to overdetermination, the possibility of experiencing oneself as the host of the other.

The equations of the epidemiologist, if they fail to take these nonfinite and deconstructible "cultural" contexts into account, will have little or nothing to tell us about "nature." The epidemiological qualities sometimes thought of as belonging to a virus, such as its reproduction or fatality ratios, do not have a directly accessible value independent of the population it is infecting and the "cultural" determinants of their behavior and health (Delamater et al.; Fuller; Cepelewicz). Nor is there a straightforward measure of number of cases, proportion of asymptomatic infections, immunity and susceptibility, or incubation period that doesn't grapple with the uncertainty of data due to vagaries of individual behavior, access to testing, reliability of government and media reporting, and so on. Thus, the expertise required to model such a viral spread or to craft policy in response crosses, reinforces, and deconstructs as many barriers as the virus "itself"; the researcher in public health or epidemiology cannot possibly study "the virus" or "the pandemic" — they have no object at the heart of their field — unless they read its traces from a field inscribed by an internally differentiated population's access to paid sick leave, health insurance and unemployment insurance, food and basic necessities, Internet, housing, safe workplaces, and organized labor, by their trust in media, medicine, and government, and anything else that might determine how a disease spreads among us.

The task of the epidemiologist is always to surmise "innate" properties by reading this context, in order to make predictions relative to what can only be another, changed context. There is no reality or nature of the virus separate from these contexts, which include even the work of the scientists attempting to understand and control them. The epidemiologist is a figure somewhat like Jonah, whose prophecies are most successful if they are heeded, and thus prove untrue.[3] Nor should the recognition of this entanglement of nature and culture be attributed to some "agency of matter" that, it seems to me, can only serve as a political alibi for those who would rather not grapple with the extent to which every seemingly "natural" given can, perhaps, be displaced or deconstructed for the sake of what, if it is not simply *the good*, nonetheless holds open its promise.

Verticality: Virus as Gene

Virology may be considered as a branch of cellular genetics.

—François Jacob, *Genetic Control of Viral Functions*

The phenomenon that led Twort to hypothesize a spontaneous, autoimmune cell product was what virologists today refer to as *lysogeny*. This term refers specifically to a "life cycle" of viruses that infect bacteria, but analogous latent infections can be found among every class of viral hosts. Most people's mental picture of a viral infection is probably closer to the *lytic* cycle, where a virion infects a cell and makes use of the cell's ability to transcribe, translate, and replicate its genes in order to rapidly generate copies of itself.[4] This leads to the lysis of the cell, its bursting and dissolution, which releases these virion particles to infect new host cells. The resulting host-virus relations fit with the average mental picture of germ warfare — such a process grows exponentially, spreading a virus horizontally through a population and leaving death in its wake (or new, viral life), unless certain cells have or develop an immunity to this infection. Ultimately, any viral process will complicate self-identity, but the roles and relations of self and other, good and bad, life and parasite, health and sickness, are more traditional in this lytic picture than what one finds in lysogeny, where no immediate viral replication takes place, but instead the viral genes lie dormant in an episome or integrate themselves into their host's genome. These genes may remain entirely unexpressed, or they may produce any number of effects that can be detrimental or beneficial to their hosts — the fates of host and virus are now tightly linked in symbiosis, and protecting its host may preserve the virus as well. Its genes will be copied when the cell reproduces, and it will "infect" this cell's offspring in a "vertical" fashion that is almost indistinguishable from the hereditary inheritance of parental genes (much like a virus, the genes we are born with can prove either beneficial or detrimental, sometimes even lethal, to us). That being said, the prophage, as this vestige of viral genetic material is called, has ways of reasserting its independence (what Koonin would doubtless refer to as its "degree of autonomy"); for example, under certain stress conditions, as if the virus were sensing that relying on its host's reproduction might not be such a sure bargain, the prophage will transition to the lytic cycle, allowing its offspring to seek a new lineage to infect.[5]

The terminology of *lysis* and *lysogeny* may be somewhat confusing given that they seem etymologically synonymous — lyso-geny would mean causing lysis. One refers to a virus (often one and the same virus) as lytic if it will cause a lytic infection in its host, lysogenic if lysogenic, but properly speaking *lysogenic* is a property of the *host*.[6] That is, the infected bacterium becomes lysogenic, meaning that it can generate lysis in uninfected bacteria. Lysogenic would literally be a modifier of the bacterium, but it has migrated to the virus by hypallage.

An eminent virologist, André Lwoff, who shared the Nobel prize with François Jacob and Jacques Monod for research into microbiological genetics, summarized the difference between the lytic and lysogenic cycles thus: "It is quite clear that a virulent phage [. . .] *behaves* as a parasite whereas the prophage *behaves* as a gene. The problem as to whether bacteriophage is or is not a parasite or a virus is entirely a matter of definition" ("Lysogeny" 322).[7] (A virulent phage is one that causes a lytic infection, whereas a prophage is the relatively latent form viral genes take during lysogeny.) In addition to destabilizing our understanding of parasite and gene (given that the "same" virus can produce both effects, often in the "same" host), lysogeny complicates the notion of immunity.[8] A cell that becomes the abode for a viral prophage often receives immunity from infection by related viruses (Villarreal 64–65).[9] Not only does it receive this immunity, but it can also cause a lytic infection in cells that lack the prophage. It passes both this immunity and this power of destruction to its offspring, creating a mark that operates somewhat like a speciation event — because of this viral infection, one lineage is of a sudden unable to live in harmony with its semblable.

In fact, viral inoculation is the best method that has been discovered to protect bacteria in industries that rely on them. For dairy manufacturers who use bacteria to make yogurt or cheese, viral infection is a financial risk — while the best protection against it is *another virus*. Infecting their stock with a harmless or defective virus provides immunity from infection (Villarreal 84–85, 170). Just as the virus *resembles* a gene, spreading by vertical descent through "natural" reproduction and marking its lineage with distinct powers that may prove beneficent or ruinous, it *resembles* an agent of immunity, the very system by which self is distinguished and protected from the intrusion of the other (Gilbert et al.).[10] Selfhood and species-identity are mediated by virality, alterity.

My primary interest in examining and deconstructing the ordinary understanding of these processes is to observe and perhaps welcome the disorder they introduce into conceptions of self and other, life and death, immunity or heredity and infection. Lwoff, who played a pivotal role in the early theorization of lysogenic processes, introduced a historical review of the scientific study of lysogeny, published in 1953, with the following: "Lysogeny occupies a privileged position at the cross roads of normal and pathological heredity, of genes and of viruses" ("Lysogeny" 270).[11] What he describes is not without relation to what Freud called "pseudo-heredity," an infection or trauma that spreads through a family tree, imitating proper heredity (209).

Shortly after what would, in hindsight, be recognized as the discovery of the virus, at a moment when, pivotally, there was not a distinct name for this object of study, early theorists of genetics made even stronger claims regarding the parallelism, filiation, or perhaps undecidability of "gene" and "virus." Hermann Muller, in an article on the theory and prospects of genetics in 1922, at a time when heredity was being studied in Mendelian terms but when little was known about its material substrate, noted the parallel between the "ultramicroscopic particles" of the genes and what he called the "d'Herelle phenomenon," the bacteriophage. He described the latter as a "phenomenon related to immunity," the agent of which, like a gene, was self-propagable or "autocatalytic."[12] Thus, "although it may really be of very different composition and work by a totally different mechanism from the genes in the chromosomes, it also fulfills our definition of a gene" (Muller 48). This question of mechanism leads Muller to a speculative question:

> That two distinct kinds of substances — the d'Herelle substances and the genes — should both possess this most remarkable property of heritable variation or "mutability," each working by a totally different mechanism, is quite conceivable, considering the complexity of protoplasm, yet it would seem a curious coincidence indeed. It would open up the possibility of *two totally different kinds of life, working by different mechanisms*. On the other hand, if these d'Herelle bodies were really genes, fundamentally like our chromosome genes, they would give us an utterly new angle from which to attack the gene problem. [. . .] It would be very rash to call these bodies genes, and yet at present we must confess that there is *no distinction known between the genes and them.* (48; emphasis added)

Muller offers a theoretical alternative for future research: either gene and virus are both what we might call Error Prone Replication systems with different material bases, thus *two totally different kinds of life*, or there is *no distinction* between them — that is, between the virus and life itself, the proper reproduction of living heredity. His intuition seems theoretically sound, and moreover seems to saturate the field of logical possibility such as it presented itself at the time. Nonetheless, neither of these alternatives are broadly accepted today — the virus uses the same material structures as the cell, nucleic acids, to mediate its heredity, yet viruses are not typically considered to be the same "kind of life" as cellular beings. They are often not considered alive at all. While they can be said to have genes or to infect us with their genes, the virus is typically not thought first and foremost as an extension of the cell and its heredity.[13] An escaped gene, perhaps, but escaped long ago, once upon a time, never to return in the same form — such that even if they re-embedded themselves in a living genome at a past date approaching the same mythic ancientness, they still are distinguished from host genes as "viral fossils" — they remain, as Koonin would say, "autonomous parasites." If they are not genes, but are not another life, what are they? What possibility was so subtle or obscure that Muller could not even identify it in theory? Well, they are *viruses*. In other words, what separates them from what we pretend is the proper essence of life, what makes life appear essentially distinguishable from the virus, is *the name virus*. And its difference from the name *life*. In the words of Lwoff, "viruses should be considered as viruses because viruses are viruses" ("Concept" 252). There will be more to say about this viro-tauto-logy.[14]

It is easy to imagine that this undecidability of "virus" and "gene" is simply the confusion of certain scientists who, in 1922 or 1953, operated in a relative fog of ignorance. And thus, that we have extricated ourselves from it. We have already seen contemporary examples that should dispel this illusion — the citation from Koonin that opened this book demonstrated that it is no easier today for a virologist to rigorously draw a line between gene and virus. It is not as simple as, for example, identifying certain nucleic acid sequences specific to viruses (or invoking the dubious notion of "self"-reproduction — on which, see chapter 4) to claim that we have passed through Muller's Scylla and Charybdis — in other words, to claim that the virus may have a genetic machinery that is compatible and even substitutable with the machinery we call living, but that we need not fold it within our essence or our family.

These biologists are sensing a risk to their family tree. From the perspective of a genetics that hopes to stand within a single and stable organic structure, they are attempting to control, contain, circumscribe, in short, to quarantine, with this nomenclature or taxonomy, something that seems to threaten that tree with disorder, disease, perhaps the end of life. If we acknowledge that the "virus" can be a "gene," everything that followed from life's "horizontal" exchanges with itself will be granted in an instant — birth will be piecemeal, with some of its particles floating freely outside the already born, befalling them long after the "origin," as if by chance, without stable hierarchy or generations.

Rather than deriving from a peculiar phenomenon or observation, the undecidability of gene and virus is what makes anything like life or science possible. Wherever one hopes to distinguish cause from effect, the virus is there already. If we think of the virus as something befalling a cell from outside, no rational thought or technological manipulation, no truth or causality will be salvageable unless one imagines something like a possibility or vulnerability that belongs to this cell from the origin. If the possibility of infection by this particular virus developed late in the course of this cell's life, that only means that at an earlier stage it possessed the possibility for this possibility, and so on until one reaches the "beginning." *On the other hand*, if viral infection is "itself" an innate possibility, if its risk or chance is intrinsic to life, it is still necessary to imagine a state of affairs inside or outside the cell that could serve as this gene-virus's elicitation. The gene must be "triggered" by the environment, etc. Gene and virus, internal and external, necessity and chance are mirror images, there is only possibility where one awaits some activation or trigger, and so the "outside" is a structural component of the system. Thus, there can be no simple answer as to whether a gene or virus belongs to the inside or outside of life. Every distribution of gene and virus, essence and accident, will be set in motion by a *virality* that leaves this structure open to the radical displacement or intrusion of *impossibilities* that can never be exhaustively reappropriated within the symmetrical and immobilizing gene-virus analogy.

Historian of science Ton van Helvoort has defined what he calls, following Ludwik Fleck, two "thought styles" that preoccupied the twentieth-century development of virology. The "exogenous" and "endogenous" thought styles framed a debate over whether the virus was essentially something external that infected the cell, or something that originated internally, from the cell itself. This question, which re-inscribes itself in the terminology and practices of successive generations of virologists, is a symptom of the undecidability of autonomy and parasitism. It is the playing out of vulnerability and infection, possibility and trigger, a logic that can account for everything, save the impossible.

The Sex Virus: Bacterial Conjugation and Viral Horizontality

I introduced lysogeny above to complicate the relations of vertical descent and viral infection. The relationship of horizontal processes among the living, such as Horizontal Gene Transfer, to virality is just as destabilizing or deconstructive. In other words, it is not the case that the virus *is* the horizontal; rather, by resembling all horizontality and all verticality, it suggests what makes them both possible and impossible. It is not *the* virus that introduces or carries out this undecidability, not some particular material structure or family of descent — which after all would only be another tree — but something like *virality,* which is no more specific to these "viruses" than it is to anything else. To the extent that even this "virality" is captured within a system that opposes it to some vitality that is there already, it serves no better than any other name to refer to the movement of impropriety and substitution, a-filiation, that is the condition of possibility and impossibility of life — the very term undergoes a process of transgressive appropriation and improper resemblance to all propriety or proper reference. *Virality* is at once a word and the thing it purports to name and can only be either by being subjected to forces that must precede it and for which it can no longer be the proper or common designation.

Joshua Lederberg's research formed one meeting point of these questions. In 1952, he authored an important review of the implications for heredity of genetic material that could be found outside the nucleus of a cell, in its surrounding cytoplasm (Sapp, *New Foundations* 157–64; Sapp, *Genes* 85–92). There was a plethora of names for this material at the time — pangenes, bioblasts, plasmagenes, plastogenes, chondriogenes, cytogenes, and proviruses — each implying a theory of its origin and function (Lederberg 403). Lederberg hoped to avoid some of the increasingly mythopoietic constructions that were forming around these characters by merely calling them all *plasmids*, a term that is still in use, though with a restricted sense. On the one hand, research into "cytoplasmic inheritance" or "extranuclear heredity" challenged the autonomy of the nucleus and its chromosomes, or the analogous genophore in bacteria, because it considered the possibility that effects of heredity could be mediated by other structures (Lederberg 403). At the same time, because certain of these structures were thought to be viral or symbiotic in origin, such research complicated the notion of heredity altogether — Lederberg called the cell the "meeting place" of "genetics, symbiotology, and virology" (426). This crossroads even led him to question the concept of "self-reproduction" (423–24). At this junction where the undecidability of vertical and horizontal, inheritance and infection, being and resemblance is most acute, Lederberg introduced the phrase "infective heredity" in a perhaps vain attempt at mediation (413). He used this phrase in particular to describe transduction, a form of Horizontal Gene Transfer mediated by a virus-like structure, which he described as "functionally, and perhaps phylogenetically, a special form of sexuality" (Lederberg 413). With respect to the question of origins, and thus of being or seeming (is this or that plasmid a gene or gene-like, viral or virus-like?) he discouraged researchers altogether from such speculation, acknowledging an artificiality and undecidability of such categories — and yet, he himself cannot write at all without implying such decisions, such as those he draws among the various "forms of plasmid," which he describes as "the hereditary parasites as against the functionally coordinated plasmagenes, with the mutualistic endosymbionts somewhere between" (Lederberg 425). He described the aim of his project as "the obliteration of semantic barriers," but one can see that, for the virologist as for the virus, the maintenance and creation of barriers is just as productive as their dissolution (Lederberg 426).

Just as vertical relations of heredity were destabilized by the undecidability of virus and gene, virality haunts the mechanisms of horizontality among the living. We can take as an example one of the primary means of Horizontal Gene Transfer, a process typically referred to as bacterial sex or conjugation. Conjugation is not directly connected to the process by which bacteria "reproduce," binary fission, which results in the production of two cells, typically genetic clones, and is considered "asexual." Conjugation, on the other hand, is described as sexual because it involves the exchange of genes between what we take to be distinct individuals. There are perhaps other analogies that suggest themselves to scientists observing conjugation — for example between the phallus and the sex pilus a donor bacterium constructs to transmit its genes (this donor is often referred to as male). From the earliest observations of conjugation to the present, relationships have been noted between the process and virality. In 1952, William Hayes hypothesized that a viral agent might mediate this process:

> It is known that symbiotic bacterial viruses can transfer hereditary characters to heterologous strains, and that *K* 12 [a strain of *E. coli*] harbours a virus which can be liberated by small doses of ultra-violet light. The known facts of recombination, and especially its marked enhancement by small sub-mutagenic doses of ultra-violet light and the presumptive one-way transfer of the genetic agent from 58-161 to *W* 677 [mutants of this strain], suggest the possibility that this agent may be a virus. ("Recombination" 119)

In 1971, another microbiologist studying the phenomenon, Charles Brinton, made a more sweeping suggestion. He argued that, as more had been learned about conjugation, it had gone from seeming "virus-like" or "analogous," to being best understood as an "essentially viral system." He based this argument, however, not on a filiation of this process from a known viral ancestor, but on analogy: he wanted to describe "bacterial sex as a virus disease and [. . .] the mating act as a virus infection." Toward this end, he proposed defining a new class of viruses that were never external to the cell and allowed the transfer of host genes, which he named "sex viruses" (Brinton 106). His suggestion never caught on. Still, it is illustrative of the analogy or resemblance that many researchers intuitively sensed between a horizontal genetic exchange and a virus. This process that resembles sex without reproduction, and thus resembles vertical inheritance without being it, also resembles a virus.

As more has been learned about bacterial conjugation, its viral aspect has become more pronounced, in a manner that complicates *resembling* a virus and *being* one. Bacterial sex also presents an analogy with sexual difference. Only bacteria who possess a certain group of genes, called the fertility factor or F factor, can engage in conjugation as a donor, and they can only give their genes to a bacterium lacking F. Moreover, the F factor is often transferred in the process of conjugation, sometimes along with genes from the donor, such that the sex role of the recipient is reversed. This F factor happens to be a plasmid, a DNA molecule within the cell but typically found separate from its chromosome or genophore. The plasmid also closely resembles the form often taken by prophages (the viral genes in a lysogenic cell).[15] Was the origin of sex a viral infection?

Several authors have explored the hypothesis that sexuality, meaning the exchange of genetic material between individuals, originated because of an infectious parasite (Hickey; Bell; Birdsell and Wills). While these authors are not particularly concerned with whether that parasite should be considered viral, their analysis of sexuality as parasitism reveals an undecidability of genetic exchange. While it is commonplace for evolutionary theorists to assume that sexuality, like everything else, must exist because it is beneficial to the organism, and thus to posit supposed advantages it would offer such as increased genetic diversity, the contagion hypothesis suggests that sexuality exists because it benefits the spread of sexuality. At least in its most primitive forms, the once sexless bacterial cell would be manipulated into sexual exchange by a genetic parasite, the plasmid, whose spread resembles that of a virus in several ways: those who have been infected by the F factor are immune to infection from other fertile or infected cells, which instead seek out only uninfected potential hosts (just as a lysogenic infection may provide immunity from homologous viruses). As a result (this is true of all "horizontal" processes), "sexuality" can spread through a population even if it is detrimental to the reproduction of its hosts. Eventually, this parasitic or selfish DNA would saturate a population, at which point selection would favor its becoming less detrimental to its host — thus it has been called "semi-parasitic" and the "fossil" of an "ancient infection" (Hickey 529; Bell 355).

The effort to determine if this "semi-parasite" is viral in nature will summon the specters that haunt every search for origins. Luis Villarreal, a virologist who argues that the evolution of life has been co-constituted in a symbiotic fashion by viruses, claims as part of this project that the bacterial fertility factor derives from a virus (80–83). Like every attempt to fix the causal chain of a proper filiation, Villarreal's quest must substitute resemblances or a-filiations for genealogical descent. The circularity of this method is manifest already in his section headings: "How phages resemble plasmids" and "How plasmids resemble phages" (Villarreal 79–80). The term *plasmid* refers to a loop or strand of DNA, separate from the chromosome in the bacterial cytoplasm, of which there may be hundreds in a given cell. A virus that causes a lysogenic infection leaves a phage in the cell that is basically identical in structure to the plasmid, though it may have genetic sequences that distinguish it. For example, many phages contain the genetic sequences necessary to create the protein casing (typical of viruses) in which it packages a copy of its genes, to travel outside a cell and infect another bacterium. Now, any question of whether a given element belongs to a certain family, say, whether "plasmids" are "viral" in origin, will depend on how we define *plasmid* and how we define *virus*. Villarreal points out that some researchers define these terms in an explicit though futile attempt to prevent any conceptual crossover or infection. The term *episome* is often reserved for prophages, for DNA of viral origin inhabiting a cell, and one distinction commonly made between episome and plasmid is that the former has the ability to integrate into the bacterial chromosome. Though the F factor is typically referred to as a plasmid, Villarreal notes several similarities it has with what are typically called episomes, including the ability to integrate itself in the chromosome.

These similarities could be decisive only if this taxonomy (the typical episome/plasmid distinctions) were absolute. In his own way, Villarreal is just as dependent on reifying an arbitrary system as those he opposes, who institute this system — it would seem — precisely to domesticate plasmids (and perhaps sexuality), and thereby to foreclose their virality. Originally, the term *episome* was introduced by François Jacob and Ellie Wollman with a purely functional definition, referring to any cytoplasmic DNA that had the ability to integrate into the bacterial chromosome (311–14). Thus, it included both viral prophages and the F factor, though only in an analogical or functional category, not suggesting common descent. Just a decade later, it was argued that this category was "artificial" and obscured the true nature of the elements it attempted to categorize, because the ability to integrate into the chromosome could be gained or lost by a particular "plasmid" or "episome," by mutation or transfer to a less compatible host (Hayes, "What"). This suggestion led to a discussion among microbiologists many of whom felt these categories could be salvaged to preserve mutual intelligibility, yet none of whom could reach mutual agreement on usage (Hayes, "What" 8–11). Today, the nomenclature persists in the sort of disarray or instability typical of virality. On the one hand, *episome* is frequently used to refer specifically to prophages, implying a viral descent. On the other hand, its functional definition (referring to the ability to integrate) is sometimes still used. Even though *plasmid* is often still defined as a purely "extra-chromosomal" element (Funnell and Phillips), the F factor is most frequently referred to as a plasmid, though one still finds occasional reference to the "F episome," in reference to its integrative capacity (Casali 35). Trying to add the F factor or anything else to the family of viruses (*or* arguing for its mere resemblance) depends on a prior unstable decision as to the essence and nature of viruses; as in every case, *virality* is not something that can simply be attributed to this or that entity but is what destabilizes (even as it makes possible) all such attributions, making the *artificiality* of the designation the most *natural* thing of all.

The problem is not merely that none of us witnessed these ancient events (such as the "origin" of sex) or that they have left few fossil traces — in fact, witnessing events never promises an understanding of their cause, which is precisely why the iterability of scientific knowledge formation is necessary, and we always determine what is a fossil of our past to some extent by deriving it from the "present." On the one hand, everything that exists is the same age, and on the other hand, there is no past and no present, nothing has a fixed age, because nothing exists within a unified time or present moment. Every fossil, in order to be a fossil, must be in some way present, even if that is only the "presence" of a trace. At the same time, for this very reason, it is impossible to settle what presents itself in that present into a representative relationship of past ages. There is only the deceptively uniform surface of this present from which to isolate representatives of the past, yet the present can only appear at all as the traces of filiative relationships, as an inheritance of forces and names. We will never decipher the origin of that inheritance without relying on analogies constructed from its traces. Even when that "past" is frozen in stone, ice, the written word, or archival documents, it is still possible to watch it change as the context of its interpretation shifts underneath it. It always remains open to *geneanalogical* exchanges.[16]

Now, if our reconstructions of the past depend on what we can intuit as related or analogous in the present, we will face an abyssal aporia when attempting to fix the target of any question of origin, such as the origin of sex, or the origin of the virus. In order to discover the origin of sex, and to decide if it was viral, we must know what sex is, and what a virus is. But if being a virus means descending from the family of viruses, we will be caught in a definitional circle that can only discover what it already knows. Analogy is no better able than genealogy to settle this inquiry, which must decide if a virus resembles another virus without knowing what a virus is. Analogy as intuition of resemblance is powerless because something that never simply appears, something like the definition, form, *eidos*, or iterability of the category, is essential to its valid analogizability — it is never simply a matter of resemblances, but of what will count as resemblance and what dis-analogies can be disregarded, a question of "essence" that exceeds intuition while founding it. Rather than simply having access to a true essentialism or genealogy that could ground this derivation, we are left to reconstitute it from traces — it must (1) take the form of history or dispersion because its law is not known in advance and no law or essence can contain it, because (2) *on the one hand*, there is only the history or genealogy of definitions, one cannot discover the essence or law outside of history but must search "within" it for what exceeds it, yet (3) *on the other hand*, for this very reason, that history or genealogy, even or especially when it takes the form of intellectual history, history of science, or history of ideas, can necessarily only start from an already formed definition or ideality, from a particular formation "within" the history it is attempting to narrate and ground. There is no means for a history of the virus or of sexuality or anything else to choose its starting point or the links in its legacy without depending on its own ungrounded and ungrounding inheritance of iterable terms. It makes our specular or speculative task no easier to recognize that these very terms, given the impropriety with which they infect our discourse and spread without regard for the essence or filiation of their hosts, these very terms or germs are viruses, even if we will never know what that is.

Villarreal, then, even as he claims to be discovering the "progenitors" of bacterial sex plasmids, is in fact searching for a peculiar sort of resemblance (83).[17] As if he were a psychoanalyst searching for the traumatic origin of a symptom, even negation becomes evidence of a repressed presence. For instance, the fertility factor lacks the genes for coding virion structural proteins (which form the shell that allows a virus to travel outside the cell and infect a new host), but Villarreal does not consider this a mark of non-belonging. We have already reflected upon the undecidability of "gene loss" in any genealogy, which lets even absence be a trace of presence or filiation. By a similar principle, Villarreal points to "defective" and "satellite" viruses, which may lack these structural proteins as well, but reproduce when another virus, or even a complementary defective virus, infects the same cell. My objective is not to discredit this analogy, as if such a thing were possible, any more than I would hope to endorse the dis-analogies emphasized by those who take a position opposed to Villarreal's. What sense is there in such opposition when the "same" datum serves at once as analogy and dis-analogy? Certainly, many of the analogies, resemblances, or similarities noted by Villarreal are quite compelling — rather than dismissing them I would simply note the effect of this necessity, that the search for origins begins from a-filiation, a resemblance that can only be determined as such by means of a secret narrative (Derrida, "White Mythology" 243). It is the similarity — in the generalized sense in which even dissimilarity counts as similarity — the *iterability* of what we call "virus" and what we call "sex plasmid" that authorizes the thesis of their filiative relationship. Rather than simply leaving us uncertain about the order of priority, this means there never has been a true origin — the very question that asks after the "original" virus or plasmid and its descendants can only be posed with a supplement of textuality, by means of something like this term *virus*, which, though it could not possibly have been there at the "beginning," allows for something like an iterable ideality and thus for a thing to appear as the origin of another thing. In this vein, I would note that Villarreal's project, understanding the virus as a symbiotic partner in the evolutionary history of life and thus as, in a sense, part of a single vital-viral family or family tree, can already be deciphered in his family name: Vi(lla)r(re)al.

This supplementarity or parasitism of the origin, which I described as a textual structure, should not be attributed to language in the traditional sense, and certainly not if this is restricted to a capability thought to belong uniquely to the human being — every living thing draws distinctions such as those that identify predator and prey (what to eat and what to flee), what is self or community and what intruder, what to mate with and thus what is its species — in short, every living thing must recognize by means of traces an origin that never was, by a method that is just as true or "scientific" and just as false, fictive, imaginative, or creative as our own. The a-filiations of the origin are recognized by a method that undoes its own authority or self-evidence.

Thus, it is not a simple matter to say whether the virus *resembles* a horizontal mediator of the living or whether it *is* one. No easier than it was to fix the virus as a substrate of verticality or something that, by imitating verticality, undermines it. The problem is just as much how the virus disorders life, how it intrudes upon "proper" genetics or filiation like a contagion or a parasite, as how it resembles life too closely, usurping or subverting the very propriety of the proper. What can be said of something that is neither gene nor germ yet exists as both at once? How can "the virus" seem, at once or in an impossible simultaneity, to grant life its powers of filiation and a-filiation, and yet to disorder or corrupt them? A response to this question can only start from the virality of its own resources (language, thought, science, being, etc.) — there may be no *virus-that-is-one*. If it is "possible" both to *be* a gene and to undermine any systematic thought of being (whose coherence is necessarily genetic), it must be because the virus, like that life it resembles all too well, *is* nothing. Nothing that could belong to the presence of being or its essential possibilities. We the "living" are faced with something that exceeds the metaphysico-scientific order of thought according to which everything that happens will have been possible, for the impossible is what does not happen. According to this thought, which is no more simply essentialist than existentialist, everything that befalls the living, including what we have nicknamed "virus," will always seem to have been, in hindsight, possible from the beginning or before the beginning. If "infection" happens, it must be because life was vulnerable to it and lacked an immunity that can only be defined in the face of this disease; thus, in a sense this "virus" would belong to life. If we find, no matter how much faith we place in the common sense of this logic, that anything we could call "life" and identify as its possibility, its genetic inheritance, is just as much undone as done by this "virus," we will be faced with something that exceeds any possible order of metaphysical or scientific coherence. This excess is not simply what corrupts the possible, rendering it undecidable and infecting it with something of its other, the impossible, but *also* what gives it the gift and exposes it to the risk of its chance, what makes even the impossible "possible." For this reason, the "virus" is no more life's other than it is life itself — it dwells within the given as the *virality* of life, the haunting of life and everything else by an alterity that is neither purely assimilable nor simply opposable to it as another essence. The "horizontality" of life is the possibility that it can, by means of its "own power" or "genetic inheritance," exceed that potentiality and corrupt or efface that inheritance "itself." Rather than being attributable to some higher power, some force of vital creativity (which would only find itself compromised in turn), it undoes any grasp that those who know life, as scientists and as the living, have over their own possibility, essence, *logos,* and formula. There is no simple family of relations or resemblances that belongs under this name. It is neither the case that something we could simply call living affects itself of its own power, nor that something simply opposed and external to it ("the virus") befalls it from the outside, but that life, virus, and everything else receives its chance and its risk from a virality that makes every proper genealogy possible and impossible.

The Fundamental Inevitability of Parasites

What we call life, and what we call virus, have descended from something that is neither vital nor viral in itself, in part because it is both. How would we decide whether the "first" replicator was gene or germ? There is no "first" to speak of until some descendent has a-filiated with it or from it. Nothing guarantees the fidelity or security of this process. For instance, if we think of "gene" as what preserves the true replication of a vital lineage, and "virus" as what disrupts it by subjecting it to an alien periodicity, *nothing will ever be simply gene or virus in itself.* Anything that receives its identity from repetition or iterability receives its "identity" from the other, from a future that has never finished arriving and that renders all distinctions of identity and difference undecidable. A "virus," like a gene, can be a benefactor or a parasite, in a fashion that makes it difficult, any longer, to know what does good to what, whose benefit is preserved by whose good deed, who is self and who other. Obviously, it is not a simple matter of delineating an inside and an outside — the virus is the *obligately intracellular*.

The questions of descent posed here, of the vertical and horizontal, lead inevitably to the question of the origins of life. This has been a fertile field for something like the speculative fiction of scientists — a fictiveness or speculativeness that is never simply opposed to the scientific. The question of origins takes a logical form that never fully escapes tautology: if we have two, cell and virus for example, either A caused B, or B caused A, or C preceded and gave rise to both. The "first" virus could have been an "escaped gene" from a fully formed cell, or it might have been an element in a pre-cellular genetic network that gave rise to both what we know as viruses and cells. Yet, the question of "origin" depends on the feigned unity that is an effect of these categories — there is no origin of the virus until "virus" has been posited as a family, and the arbitrariness of this category attaches to any distinction between the "proto-virus" and its offspring, the "first" virus. My objective is not to discredit this pastime, which can produce intuitive insights and even testable hypotheses for the experimenter, nor do I intend to adjudicate among the contending narratives that have been put forward. I only want to point out that the most rigorous logic, reason, and science, which demands this search for origins, at the same time undermines anything like a coherent grounding of its descendants. In short, no matter where one looks, no matter how far back one forces imagination or speculation, one finds the différance there already. Virality, vitality. Not as two opposed and opposable elements but caught in an undecidable embrace. Derrida used this term *différance*, among other things, to think alterity without oppositional stability or simple origins — thus, if life and virus, self and other, reproduction and contagion seem to form opposed families, there is no original ground that would bestow unity on these categories, and thus wherever we think we meet with one or the other, it is only the provisional deferral of its "opposite," which haunts it without proper abode. We would like to find the first — virus or life — and yet we still do not know what that is. Then, it is only by projecting the past in the image of a future that has not yet arrived, that we claim a "proto-virus" one fine day brought forth a "virus," or a "proto-cell" brought forth a "cell", as if these were unified categories. Every descent is a traduction or a-filiation, which is most *viral* or *virus-like* because it places all being and seeming in question.

One of the objectives of Eugene Koonin's research into comparative genomics is to approach this question of the origin of life. The story he tells is one of a *Virus World*. Shortly after its discovery, the virus was thought, because of its relative size, to have preceded life and to have been a mediator between the inanimate and animate worlds. However, once the parasitic or symbiotic nature of viruses was better understood, this seemed a violation of logic — how could the obligate intracellular parasite preexist the cell? Nonetheless, Koonin observes that any replication system is vulnerable to being exploited by parasites. A pre-cellular replication system, for instance a pool of naked (like Adam and Eve), self-replicating nucleic acids not enclosed in cell-like membranes, could easily be exploited by other, "opportunistic" molecules. This would mean that molecules capable of replicating themselves were being subverted by molecules that lacked this self-reproductive capacity but could use the productive capacity of their neighbors to get themselves copied. In Koonin's vision, these parasitic elements are the true engine of the evolution of both life and virus — the cell membrane originates as a defense against them. They are the origin of life, yet they are, in his words, "virus-like."

He accompanies this creation story with a short excursus entitled "The fundamental inevitability of parasites" (Koonin, *Logic* 320–22). *Every replication system*, Koonin explains, will necessarily be occupied by parasites, will even bring them forth of its own power. His argument for this, backed by experiment, is that any system with the power of "self-reproduction" can give rise to elements that are more efficient replicators precisely because they lack this power but can exploit others' reproductive capacity. As if selfhood itself were an otiose encumbrance. He illustrates this with a simple, schematic diagram (see figure 2.1).

Yet another tree: out of the unified stem of "primordial replicators" branches the replicators *tout court* — this simplicity apparently being a product of something more fundamental — and the "replication parasites," those who lack the capacity for "self"-reproduction but exploit the power of another. While certain precautions should be taken with its suggestion of temporal progress and simple division, I take a perhaps parasitic pleasure in this figure and the opening it provides for my possibly (who can say?) alien purposes. It represents the common ground of Koonin's project and my "own," which are not simply opposed. I too am trying to think life and virus from the perspective of something like a primordial replicability, an iterability. However, this iterability is not something that only existed one fine day, long ago, in a warm little pond, but continues to haunt every production of apparent autonomy or heteronomy, and has never existed as itself, in a pure unity of self-identity. Like this something that is still within and without us today, which is neither virus nor gene but gives both their chance by haunting everything that seems most our own with the continual threat of subversion, and by making even the most repulsive other our future

Figure 2.1. Differentiation of a population of evolving replicators into hosts and parasites. *Source*: *Logic of Chance*, 321. Reproduced with permission of Eugene Koonin.

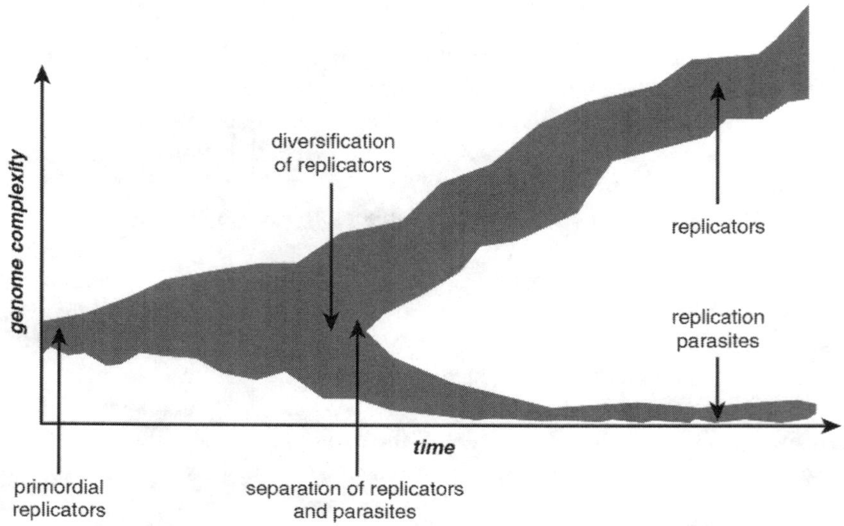

lifeblood, this iterability brings forth all virality and all vitality because it is both and neither in itself.

Nor should we think, because I have privileged certain examples related to bacteria and their viruses, that the lineage or branch of animals on which we take refuge has somehow extracted itself from this undecidability. We are sometimes imagined to have perfectly sequestered our germ line, perfectly protected genotype from phenotype or essence from existence, so that our reproductions would truly take the form of a tree. Nonetheless, the human genome is littered with remnants of viral intrusions, fossils or sur-vivals of past infections. Many viruses have the ability to insert themselves into the chromosomes of the cells they infect — these Endogenous Viral Elements (EVEs) represent instances when a virus integrated its genes into a germ-line cell and thus was able to become part of the heredity of the human race, for as long as we both shall live. Not only that, but these EVEs often encode sequences to copy and reinsert themselves into the chromosomes of dividing cells, so that they have multiplied rapidly in our genome. While current estimates tend to count a little more than twenty thousand protein-coding sequences in our genome as authentic human genes, our genome is host to hundreds of thousands of EVEs (Mayer et al.; Willyard).[18]

We could not reproduce at all unless we reproduced with these viruses and as these viruses. Many who study them start from the assumption of an antagonism between us and them, life and virus, and focus on the work our cells do to minimize their influence. For instance, we "defend" against EVEs by silencing the expression of many of the numerous copies they leave in our cells — nonetheless, even this apparent defense mechanism must necessarily be a "positive" part of the life it is inseparable from. There would be no life or reproduction otherwise. In other words, the silencing of these genes is not the return to some mythical null or essential state, but necessarily produces side-effects, cross-contaminations, bystanders, such as nearby genes that would otherwise be expressed but are silenced because of their proximity — whatever life *is there*, whatever it becomes, will have been an effect of these viruses even in their apparent silence. Nor is that their only influence. Not every EVE is silenced. As more has been learned and more biases overcome by those who study them, EVEs have been found to be essential to a host of *vital* functions.[19] I am most interested in one in particular. A newly forming embryo requires a sort of filter that can allow nutrients to pass into it from its mother's blood, but that blocks the mother's white blood cells. The mother's immune system might view the child's cells (which are genetically different) as "foreign," and might attack them. In order to create this filter, a syncytotrophoblast is created, a fusion of several cells, which allows for the passage of the good, while preventing the risk of autoimmunity (Mi et al.). Such syncytia are often created by viruses, who obviously have their own reasons for needing this protection from host immunity. Nonetheless, we have recruited their genes to perform this function ourselves, and precisely to ensure our own reproduction. Viral genes perform a similar function throughout the class of mammals (a class defined by certain features of its maternity). Our reproduction depends on the virus.

The EVE is the end of day. It is a twilight in which no pure life is possible or visible, but everything resembles its other. Nonetheless, EVE is not just any mother, but the mother of mothers, the mother of maternity. Even if this EVE haunts every apparent life or virus, and robs from their trees, without her there would be nothing. No chance and no risk, not simply of life, but of what gives place to life by allowing something to stand over its abyss, by exceeding what has merely been possible as life's reproduction. Open to sur-vival or living on, the EVE is always the day before, the advent.

Chapter 3

Viral Origins

Antagoniste des bacilles

The virus is a concept.

— André Lwoff, "The Concept of the Virus"

This chapter takes a seemingly simple question about the history of virology, and the impossibility of arriving at a simple answer to it, as a symptom of the instabilities of virality. When was "the virus" first discovered? This question continues to provoke controversy among contemporary virologists and historians of science. Its uncertainty or undecidability is not simply due to historical accident (multiple researchers reaching similar conclusions at similar times), but is an effect of a necessary aporia in the concept of origins. At no point in its history or present has what we today call "the virus" been anything but a question for us; as a result, it is never straightforwardly known or unknown, but effects of its presence or absence, and our knowledge or ignorance, are produced retrospectively.

I focus on two moments of "discovery" that have given rise to analogous debates over priority. First, in the last decade of the nineteenth century, two researchers made similar observations of tobacco mosaic disease that are often considered to be the discovery of the virus tout court. In the 1910s, a strikingly similar situation took place between two researchers who investigated what are now understood to be bacteriophages, *viruses that infect bacteria. In both cases, the conceptual and technological framework of experimentation partly explains these converging "discoveries." The* germ theory *of disease argued that every infection was caused by a microbe, and that the pathogen ought to be culturable (samples from an infected patient should grow as colonies in the lab and be able to infect a healthy individual). Moreover, the* Pasteur-Chamberland filter, *which was able to remove everything visible by the microscopes available at that time, was thought to produce sterile samples. What to make, then, of a pathogen that could not be grown in culture, retained its infectivity after passing through the finest filters, and was beyond the range of visibility of the light microscope?*

The existence of this techno-conceptual framework explains why multiple researchers would arrive at similar observations around the same time but does not fully constrain how those observations will be interpreted. In each case, these interpretations can strike us at once as foresight and atavism — I emphasize in particular the role that shifting linguistic usage has played in inclining contemporary readers to hear their own understanding reflected in the past. Rather than simply measuring the past against our present, by pretending that what we call "the virus" is a coherent and stable category, we can use these effects of anachrony to view our own knowledge as a reading of traces. Not the recuperation of a simple origin, but the instability mobilized by that origin's never-having-been.

I contrast this deconstruction with the understanding of science-as-translation found in the work of Michel Serres and Bruno Latour. Both authors imagine it would be possible for something such as "the microbes" to communicate themselves to us directly, without the intermediary of reading or the virality to which it remains exposed.

***

The tree-structure, which is the very principle of establishing logical or genealogical relationships, is co-constituted by forces that have no place or name within its system, but which manifest only as its fictionality, the displacements that subvert its systematicity. Moreover, rather than resulting in something more true, more stable, or more systematic than any tree, this transplantation can only result in yet another similarly fragile and fabulative tree-structure, whose fictionality has the force of natural necessity. What we call "the virus," then, is no more rigorously categorized as a horizontal intruder relative to the tree of life than as the very agent establishing life's filiative, vertical relationships. It is neither because it resembles both to the point of revealing the instability of all such classificatory structures, including those that give "viruses" their provisional intelligibility. At the end of the previous chapter, a rethinking of the value and position of origins followed from this questioning of the rootedness of our trees. Something that could only arrive long after any origin, which in fact remains yet to come, nonetheless co-constitutes originality. A future that has never finished arriving will determine what a "virus" can be or become, and what emerges from this punctured horizon determines what can be counted as the origin.

Moreover, because of this structure of necessary non-closure, the work of the scientist, which always involves establishing knowledge of causes or essentialities, in short, origins, has a supplementary relationship to the subject matter it studies. *Supplement* refers to what is at once necessary to the completeness of a system and yet exceeds its boundaries, revealing its necessary incompleteness. There is no scientific concept — term or germ — such as "virus" or any other, that is not a provisional and artificial (*naturally* artificial) supplement to scientific investigation. Yet, there is no way to return to the "origin" or the "essence" — of what? Of a "virus?" What is that? — without such supplementation. The belatedness of originality is no more objective than subjective, no more belonging to "nature" than to "science," but is the uprootedness inhabiting "both," which their division and opposition attempts to stabilize. We could refer to this, in psychoanalytic terms, as the *nachträglichkeit* of origin, its dependence on what seems to follow it and even derive from it. Now, lacking originality, not belonging to the family tree, depending on and feasting on what one is not and cannot be, is the status of the parasite. This supplementarity is the parasitism or virality of origin, a parasitism that is the origin of originarity "itself." It creates a strange mirroring between a science such as virology and its objects, what we call viruses. Virology, having a parasitic relationship to the origin it co-constitutes, is itself a virus, an object that falls within its domain of study.

To explore this virality of virology, I will examine certain debates or insecurities that persist today surrounding the "discovery" of the virus. In the 1890s, two researchers made similar observations of an unusual pathogen, and the argument over priority for the discovery that began in their lifetime persists a century later, among virologists who would like to date the birth or origin of their science, and to elect its father or patriarch. A remarkably similar priority dispute took place two decades later, between two scientists who made similar observations of what we now call *bacteriophage,* but who interpreted their findings quite differently. These intractable disputes reveal that the origin (the "first" virologist, the first to recognize a virus as virus) never was. This type of priority dispute, which is not at all uncommon in the history of science, has sometimes been referred to by historians and sociologists of science as *multiple discovery*. This term tends to be associated with a particular interpretation of the nature of scientific discovery; it would question the role of individual creativity, genius, or psychology, by arguing that discovery's simultaneous emergence from several researchers demonstrates an almost deterministic scientific progress that depends on historical circumstance or context (Ogburn and Thomas; Merton). Thus, in the case of the "discovery" of the virus, a certain conceptual or theoretical background and certain laboratory practices seem to make the time "ripe" for multiple discovery. Germ theory had led to the expectation that the cause of infectious disease would be a microbe that could be grown in pure culture (one of a set of rules for determining the cause of illness known as Koch's postulates). The researchers investigating tobacco mosaic disease could not grow the pathogen in pure culture, because it only multiplied in the presence of healthy cells. Moreover, a device known as the Pasteur-Chamberland filter was used to produce what were thought to be sterile samples, because it was fine enough to remove any microbes visible with the best microscopes available at that time. These pathogens could pass through these filters and retain their infective power, though they remained invisible to researchers except for the symptoms of the disease they caused. Thus, the technoconceptual matrix in which this research was conducted produced a kind of dividing line that led *almost* of itself to an unexpected set of observations that seemed, to some, to demarcate a new class of entity.

Many sociologists of science questioned the theory of multiple discovery because it failed to account for how the concept of discovery actually functioned among scientists themselves (Brannigan 46–62). Regardless of how ubiquitous multiple discovery was or wasn't, the purpose of discovery narratives for practicing scientists was to solidify systems and networks of research by creating a back story that validated them. Sociologist of science Simon Schaffer refers to this process of the sedimentation of discovery narratives in terms that are similar to those I just used to describe the virality of discovery: he sees negotiations over discovery narratives as inherently "retrospective," as a "secondary reification," a version of the past necessarily "rewritten" in terms of the present, in order to affirm the scientific culture in which these narratives emerge ("Scientific" 397; "Making"). These negotiations and renegotiations at the same time revise notions of scientific authorship and authority, displacing not only our representations of scientific subject matter, but of the proper and authoritative modes and technologies of its publication and dissemination (Csiszar 159–98). If there is something like a deconstruction of science *on* which I am writing here, it is not simply opposed to this sociology of science and the negotiations with or rewritings of the origin that it describes. Nonetheless, the scientific "communities," "theories," or "practices" that play a role in these sociological narratives or reconstructions are just as much riven and undermined by the narratology that grants them their only chance. However stable and recognizable such historical formations appear in a given light, they rest on the non-foundation of a reading or rewriting without origin prior to it. In what follows, I attempt to narrate one or several scenes of viral discovery as if from the perspective of a generalized undecidability, dividing every term or germ from itself, and making every appropriation and reappropriation of its forces of discovery and ignorance possible and impossible. What monstrous family might grow from such roots or such soil?

The Word Virus

If we would like to return to the *discovery of the virus*, we will always arrive too late. Every "discovery" fails to recover a pure origin, to find something that has not already been sussed out and infected, a discovery *of* the virus, double genitive. If one turns to those texts that are typically put forward as the threshold of this viral discovery, the most immediate sign of some *contretemps* or anachronism is the ready designation — as if it were always already there — of the virus *as virus*. There was no need for these investigators or writers to announce that they were coining a neologism or even granting it a specialized use; it was foreordained that what they were investigating was a *virus*, which at the time could refer to any infectious cause of disease. Even in its early life in Latin, *virus* could refer to a venom or poison, such that one can trace these effects of anachrony back to the first century CE, when Aulus Cornelius Celsus wrote of dog bites in his *De Medicina*, "Especially if the dog is rabid, the virus must be drawn out with a cupping glass" (qtd. in Hughes 109; Waterson and Wilkinson 3).[1] There was no contradiction, in the late nineteenth or early twentieth centuries, in speaking of a *bacterial virus* — meaning not a virus that infects bacteria, but a bacterial agent of disease.[2] (One vestige of this non-specific or generalized usage is our word *virulent,* which can still be used to refer to the severity of any pathogen.) M. W. Beijerinck, investigating a mosaic disease of the tobacco plant in 1898, needed no hesitation or justification to write, for example, "reproduction takes place only when the virus is connected with the living and growing protoplasm of the host plant" ("Concerning" 39).[3]

In short, I am concerned here with what, borrowing a phrase from Beijerinck, we could call the *Übertragung des Contagiums* ("Ueber" 19). This German word, *Übertragung,* can refer to all manners of transfer or transmission, including infection, while also referring to what we in English pretend to distinguish as *translation.*

The foremost sign for these virologists, who were perhaps not purely *avant la lettre,* that they were nonetheless dealing with something novel was the filterability of their pathogen. Here we have an image of the technicity or prosthesis of origin: the virus first appears, invisibly, as what exceeds certain conceptual and technological limits — the circuit of germ theory, Pasteur-Chamberland filter, and microscope faced its limits in confronting this virus, and yet the virus could have no definition whatsoever without the very limits it fictionalized.

The Pasteur-Chamberland filter was able to capture all known bacteria, which is to say anything large enough to be seen by the microscopes available at that time. Samples from a plant infected with tobacco mosaic could be filtered and still retain the power to infect another plant (which was the only visible sign that the contagion remained). As a result, these pathogens (several more were discovered in the coming years) were dubbed filterable viruses, sometimes called ultramicroscopic or simply ultraviruses, or inframicrobes (Lwoff, "Lysogeny" 274; Lwoff, "Concept" 241). The *filterable* designation may be a source of confusion — this word is a contranym, which can refer both to something removed by a filter and, as in this case, something that passes freely through it. According to later reflections of the virologist André Lwoff, "For obvious reasons, none of the scientists studying filterable infectious agents was pretentious or modest enough to describe himself [*sic*] as an ultra-virologist or as an infra-microbiologist. And as everybody has to be labelled, these scientists were labelled as virologists. Quite naturally, as a result of the principle of the least effort, the ultraviruses, the filterable invisible infectious agents studied by virologists, became viruses" ("Concept" 241).[4] It was not until at least this semantic shift that the question of the "discovery" of the virus could be posed with anything like the meaning such a question would have for us today.

From our vantage, with something like a modern concept of "virus" in place (as if there could ever be just one), does Beijerinck's role as discoverer become clear? Unsurprisingly, one finds as much prescience as anachronism in his 1898 text. On the one hand, he correctly (by our standards) and rather brilliantly surmises that the "virus" reproduces within the cells of the plant:

> In any case, it is reasonably certain that the virus [*Virus*] in the plant is capable of reproduction and infection only when it occurs in cell tissues that are dividing [. . .]. Without being able to grow independently [*selbständig*], it is drawn into the growth of the dividing cells and here increased to a great degree without losing in any way its own individuality in the process. In conformity with this, no ability of reproduction outside the plant could be proved. (Beijerinck, "Concerning" 38–39)

On the other hand, he declares this "virus" to be a *Contagium vivum fluidum,* a living fluid contagion, suggesting that the infectious agent actually *dissolves* in water.[5] The experiments he puts forward as evidence of this demonstrate that, not only does the virus fail to grow on agar plates, but it seeps into the agar as well. He concludes, "The experiment was, therefore, decisive in determining whether the contagium was actually capable of diffusion and, accordingly, had to be considered as soluble in water, or if not capable of diffusion, therefore, as extremely minutely distributed, yet as corpuscular, that is, as *contagium fixum.* [. . .] It seems nevertheless proved that the virus must really be regarded as liquid or soluble and not as corpuscular" (35–36).[6] This idea of a living fluid as a cause of infection carried with it the foul air of outmoded theories of disease (Waterson and Wilkinson 28; Claverie and Abergel 90).

Any question about the "priority" of Beijerinck's discovery must reckon what he thought he had discovered, against whatever it is we think he discovered. Or what we think he thought against what we think he found — my objective is to hold in suspense the framing that treats this difference as an effect of historical hindsight or of the perceptions of a fallible subject imposed on a self-identical reality. All this is possible and necessary because "virus" is, from the "first," something that must be *read.* It would customarily be said today that Beijerinck was correct to claim he had isolated a contagion that reproduced inside a living cell, but incorrect that it was soluble. Moreover, he was not the first to study this disease or even to recognize that its agent was filterable. Dmitrii Ivanowski, a Russian botanist, published a study six years earlier, "Concerning the Mosaic Disease of the Tobacco Plant."[7] While Ivanowski noted that the contagion was filterable, this did not shake his "conviction that the mosaic disease is caused by bacteria" (29). Even after Beijerinck's publication he continued attempting to grow the virus in cultures on agar plates (as one could many bacteria) (Bos, "Embryonic" 613–14). From our perspective, arbitrating these conflicting views, Ivanowski was *wrong* in thinking that a virus could reproduce outside its host, but *correct* that it was not essentially liquid. This is not to say that his culturing experiments would ever have brought him close to the understanding of viral replication that Beijerinck intuited, but Ivanowski at least avoided one of his successor's speculative errors.

I have no intention of settling the question of priority, nor do I think it is possible. Not merely in this case but in principle, and for reasons that have to do with the textuality or para-citation of the origin, the necessity that it be read. For instance, there is no way to measure the extent to which our arbitration of this debate is informed by the anachronistic or *nachträglich* force of the word or term *virus* and its semantic drift or dissemination. Our conviction that a *virus* is something essentially irreconcilable with a *bacterium* must descend in part from our familiarity with these two vocables. That is not to indulge in some sort of linguistic determinism, since after all we are capable of recognizing two distinct words within a single group of letters, but to observe the non-determinism that overtakes both what we think of as the linguistic and the non-linguistic. Even if Beijerinck's ideas about the water-soluble nature of the virus were fanciful, the fact that he saw that nature as distinct from other microbes, and distinctly parasitic, makes him seem closer to our view or our linguistic usage. Ivanowski's failure to distinguish the filterable virus from bacteria could perhaps (how would one measure it exactly?) seem less categorically mistaken to us if our usage of the terms *virus* and *bacterium* still overlapped.

More than one recent virologist or historian has been seduced by the a-filiative drift of these words. Those who attempt to decide the question of priority in Beijerinck's favor often invoke his use of the term *virus*. Virologist Lute Bos's claim, in 1995, that Beijerinck "coin[ed] the term 'virus' for a new class of pathogens" seems rather anachronistic, given that Beijerinck was applying — and certainly not "coining" — the term generally (as we would *contagion*) ("Embryonic" 618).[8]

This viral misreading was part of a debate triggered by celebrations in the 1990s of the "centennial" of virology — that is, by the implicit or explicit attempt to date the origin of virology. It is precisely among experts that questions of origin become most conflictual and least easily settled. There is a good chance that if we asked a stranger on the street, they would not have been planning to celebrate the centennial of virology and would not know nor care whether it should be observed in 1992 or 1998. On the other hand, it is precisely those whose life work involves analyzing and ultimately cathecting this field, and whose professional and personal identities are staked on these distributions of its forces, for whom such questions become particularly insoluble. This particular question became a source of contention among virologists after

Alice Lustig and Arnold J. Levine published "One Hundred Years of Virology" in 1992. They concede that "it is often difficult to assign a single date to the discovery of viruses," but grant credit to Ivanowski. They choose their words carefully, emphasizing that he was the first to report the virus was "filterable," and that "the term filterable agent was the name used to describe these organisms well before the term viruses was specifically applied to them" (Lustig and Levine 4629). Perhaps I have a parasitic ear, but it is hard not to hear the disseminative resonances of their praise when they speak of "the field of virology that he fathered" (Lustig and Levine 4631).

This article drew the attention and ire of another virologist, Lute Bos, who wrote several responses. His first, "The Embryonic Beginning of Virology," was published in 1995, followed by another with the para-citational title, "One Hundred Years of Virology?" He poses what he calls a "decisive question": "whether in 1992 virology could be judged to have *started 100 years ago*" (Bos, "Embryonic" 614). Bos emphasizes that Ivanowski persevered in thinking he would be able to culture the pathogen and thus that it reproduced outside of the plant (though he acknowledges that Ivanowksi was closer to the truth than Beijerinck regarding the particulate nature of the virus). Somewhat surprisingly, Bos concludes by acknowledging, "Hence, it is not very fertile to debate the issue who was first in discovering a virus. None of the pioneers could have known what he was talking about" ("Embryonic" 619). Still, he claims that Beijerinck viewed his findings in the right "spirit." It is worth noting, when considering these questions of legacy and inheritance, or indeed a-filiation, that both Beijerinck and Bos were Netherlanders (Horzinek, "Editorial").[9]

Scientific terms, like any other, necessarily outlive or outgrow the contexts of their creation (otherwise they would be useless for communication and writing would be impossible) — a case such as the present, where the semantic shift is relatively measurable, is perhaps a controlled and limited circumstance within a generalized dissemination that takes place always and everywhere without the earth opening up to swallow us and typically without apparent miscommunication. Which means, in turn, that no matter how rigorous our scientific understanding of this or any field becomes, the textuality of its borders and domain will continue to haunt and displace its conclusions, a textuality not at all limited to language in the traditional sense, but including every element of context that shapes this discourse precisely by being excluded from it (institutions, politics, the subject position of the researcher, sexual desire, etc.). The most obvious resolution of this relatively innocuous debate or any other alternative, even one with life-or-death stakes, facing a researcher may always shift imperceptibly.

The d'Herelle Phenomenon

Just two decades later, an analogous conflict would arise over the discovery of what we today know as the bacteriophage, a term for viruses that infect bacteria. Félix d'Herelle is often credited with their discovery, for his somewhat sensationalist 1917 paper "Sur un microbe invisible antagoniste des bacilles dysentériques" ("On an invisible microbe antagonistic to dysentery bacilli"). Its brief three pages offer a surprisingly grandiose view of its subject, some of which appears today as legitimate insight and some as ungrounded overreach. D'Herelle had cultured bacteria from the stool of dysentery patients and found that transparent plaques of dead bacteria would appear among the cultures. He attributed these *"taches vierges"* (clear or virginal spots) to the action of a deadly and infectious "anti microbe" (Summers 52). Like Beijerinck, he found he was dealing with something that retained an infectious power after passing through the finest filters. Among his speculations that we would today discredit, he concluded that his bacteriophage was the "true immunity microbe" — he did not spell out this theory in this short report on his "discovery," but it is clear that he already had in mind the theory of phage-mediated immunity that would guide much of the remainder of his life's research:

> In summary, I found in certain dysenteria convalescents that the disappearance of the dysenteric bacillus coincides with the appearance of an invisible microbe with antagonistic properties against the pathogenic bacillus. This microbe, *a true immunity microbe* [*véritable microbe d'immunité*], is an obligate bacteriophage; it is a strictly specific parasite, but, if it is limited to one species at a given moment, it can act later on different germs by habituation. *It seems thus that in bacillary dysentery, besides an analogous antitonic immunity which derives directly from the organism of the patient, there exists a heterologous immunity caused by an antagonistic microorganism.*[10] It is probable that this phenomenon is not specific to dysentery, but is of a more general order since I could make similar observations in two cases of paratyphoid fever. (d'Herelle, "Invisible Microbe" 3; emphasis added)

The future course of d'Herelle's work made clear that he thought he had discovered, not a peculiar case of phage-mediated immunity, but one of immunity's primary mechanisms (*Bacteriophage* 168–69). He theorized that all recovery and immunity from disease was a product of coinfection by such bacteriophages that made the body's own immunity, if there was any, supernumerary. He imagined an immunity that could spread through a population like an infection. He went on to extensively research phage therapies as a response to epidemics, and even attacked Pasteur's theory of immunity and the vaccines it produced.

As always, there is no absolute perspective from which to judge d'Herelle's contribution, its originality, or correctness. While his views on a body's own immunity and the efficacy of vaccines are, according to our best medical knowledge (which I am not placing in question), wrong, his innovations in phage therapy have undergone a radical reevaluation in the past few decades. Some of the disillusionment in phage therapy may be attributable to poor research design on d'Herelle's part and excessive self-promotion, but interest primarily waned due to the success of antibiotics beginning in the 1940s. Unfortunately, this led to their overuse (including their overprescription for minor ailments and their use to fatten livestock), which has in turn given rise to widespread antibiotic-resistant infections (Ventola). As a result, there has been a steady increase in phage therapy research since the 1980s (Kuchment; Gordillo Altamirano and Barr). Had I been writing this narrative a few decades ago, phage therapy would have seemed another of d'Herelle's crackpot ideas, rather than a prescient foreshadowing of modern medicine.[11] My retelling of these events is not free of any of the viral or anachronistic effects of text and context that I am trying to adumbrate; I can, as much as possible, try to mark or re-mark their enabling constraints, but certainly cannot step outside them.

By no means does this renaissance of phage therapy vindicate d'Herelle's view of the immune system, but it demonstrates that productive conclusions can be drawn from false or exaggerated premises. Not only that, but both "premises" and "conclusions" depend on a shifting frame of reference that allows for their evaluation, and within which the values of truth or falsehood, good or bad, efficacy or inefficacy, causation and correlation, and others are inscribed. The deconstructibility of this frame is the only given, which makes truth depend on something that can no longer simply be called fictional, given that there is no independent truth it can be opposed to.

None of this is what d'Herelle is primarily remembered for today: the "discovery" of what he called an "obligate bacteriophage" ("Invisible Microbe" 3). He coined this term, which etymologically means *eater of bacteria.* As with Beijerinck, part of the rationale for crediting d'Herelle with priority involves his identification of the virus as something essentially external to what it was infecting and parasitic on it. The context of "discovery" also paralleled Beijerinck's in that d'Herelle was preceded by another investigator who had observed more or less the same phenomenon — and perhaps shown more caution in theorizing it. I quoted in the previous chapter Frederick Twort's observations, published in 1915, of transparent plaques forming in colonies of bacteria, much like those observed by d'Herelle. Twort also noted that bacteria preserved from exposure to this contagion, and which bred for several generations without any appearance of disease, would nonetheless "spontaneously" form these epidemics. He described the bacteria's "apparent spontaneous production of a self-destroying material," and compared it to the formation of cancer (Twort 1242). The risk of acknowledging Twort's discovery would be that of recognizing the virus as something belonging to life itself, like an organelle or gene.

Before Twort was brought back into the debate with d'Herelle, his work was rediscovered by a French microbiologist and immunologist, Jules Bordet, who had taken issue with d'Herelle's theories. Bordet was director of the Pasteur Institute in Brussels, and by the time of his controversy with d'Herelle had won the Nobel prize for immunological research that conflicted with d'Herelle's vision of phage-mediated immunity. Together with Mihai Ciuca, Bordet published papers in 1920 challenging d'Herelle's interpretations by suggesting an "endogenous" theory of the observed bacteriolysis (van Helvoort 98).[12] Like Twort, Bordet and Ciuca thought that something produced by the bacteria itself caused its lysis (they hypothesized it was an enzyme) and led to the production of this same agent in exposed bacteria. Somewhat fatefully, they referred to this as the bacteria's "lysogenic" property (Bordet and Ciuca 1297) — a common term in immunology at this time referring to an agent capable of lysing a cell (Summers 64–66). Perhaps to undercut some of d'Herelle's authority, they also emphasized Twort's findings, and the priority of his discovery.[13]

We are today in a position to recognize that both and neither of the parties to this dispute were correct. Twort and Bordet were describing something like what we now know as the lysogenic cycle of a bacteriophage, and d'Herelle was describing something like its lytic cycle. Both stood just as much in error as in the truth. That is, phage or virus "in itself" is neither endogenous nor exogenous, neither self nor other. It is none of these discrete possibilities because it is all of them at once — potentially arriving to a cell or organism from outside or potentially transmitted as its "innate" heredity over countless generations, and capable at any point in this history of taking on the appearance of protector or destroyer. The virus is undecidably threat and promise, immunity and disease. This "exogenous" event can become the very gene of a proper heredity because there is a *virality* more fundamental than any life or virus, which gives both their chance only by letting them descend from a reappropriation of what belongs to nothing, even itself.

D'Herelle would not live to see his theory both vindicated and contradicted by a model that could explain these observations as lysogenic and lytic cycles of a nonetheless single contagion. On his deathbed, he was shown the first microscopic images of phage (which were captured in 1940 after the invention of the electron microscope), and though he probably died happy having seen this portrait of his legates, such an image can never ultimately answer the question of whether these figures are essentially internal or external to the cell (Ackermann). The only incontestable contribution of his legacy was *the name bacteriophage*. In the estimation of one historian: "Many workers had accepted the independent discovery of the same phenomenon and simply called it the 'Twort-d'Herelle phenomenon,' which gradually gave way to the simpler term 'bacteriophage phenomenon' or 'bacteriophage.' D'Herelle had, in a sense, won by virtue of giving the phenomenon a name that caught on" (Summers 77). D'Herelle's legacy: *bacteriophage* spreads like an infection, goes viral. Just as the semantic drift of *virus* played a role in the retrospective self-evidence of Beijerinck's discovery, it is our reception of the term or germ *bacteriophage,* without which no question of its origins can be posed, that fixes d'Herelle's tautologous priority. We have become its breeding ground.

Discovery as Discission

There is no science, nor any other discourse or form of knowledge, without this virality. There is no field in which or to which science properly belongs, where it can progress according to an autonomous method or speak in a voice that would be its own. Any author or text who would like to speak in its name must occupy a field and borrow resources that are already polarized by semantic and axiological values that may seem to have arrived from elsewhere yet cannot simply be called *unscientific*. Whether a given field is considered scientific or not, and whatever the guiding sense of scientificity within or against which it is so judged, it shares in the "essential" nature of science, which is to merely simulate essentiality, to take form around divisions that no single author or voice can univocally control, and which in fact undermine anything like the unity of authority.

This is just as true in the case where a term is borrowed from the tradition, like Beijerinck's *virus*, as it is when d'Herelle apparently invents his *bacteriophage*. Any such neologism depends on the preexisting system for its recognition, acceptance, and reproduction, not only in the fashion manifest in this case where etymological elements have been cobbled together, but also in the under- or over-lying sense that only allows this term (whatever form it takes) to function within a system of values such as internal/external, self/other, immunity/disease, reproduction/parasitism, life/death, and so on. This is the *virality of origin*, the impossibility of locating a simple or punctual instance in which the first virus or its first discovery (or anything else) took place — nor can this alterity or imbrication simply be parsed as a temporal separation (for instance, of a "discovery" followed by its "justification" or recognition by a community of researchers) but necessarily haunts every moment of discovery or rediscovery as its "internal" other, a dependence on something already gone and yet to arrive that nevertheless never has and never will simply expose or exhaust itself. This is not to say that nothing "new" happens, and certainly not that there is some unified and singular system that programs and executes its own future. Rather, this system gives itself to be read — and leaves an opening for an at times barely legible reorientation — for the "same" reason that there is system at all, and the iterable ideality that implies. That is, only to the extent that the event or advent of something unforeseeable "within" the system interrupts or disrupts its autonomous functioning at every moment, *as* the irruption of every moment.

Despite certain appearances, what I am tracing or retracing here, which is neither narrativity as such nor simply its impossibility, is not reducible to what Latour has called "Actor Network Theory" (ANT). Latour was critical of both classical scientific historiography or hagiography, which elected great thinkers (such as Pasteur) from whom scientific progress was said to issue forth like Athena from the head of Zeus, and of those sociological theories introduced at the beginning of this chapter, which seemed to attribute the same power of creation to a historical moment or social context. Latour aspired for a theory that could recover the role of "the agents themselves," not just the many scientists and their laboratory setups or publishing industries but the viruses or microbes — among other things — with which they collaborated to achieve their aims (for this reason, Latour has been more easily enlisted as an ally by contemporary "new materialism"). A deconstruction of science would destabilize everything in Latour's historiography and theorization that remains oriented toward a telos of "the things themselves":[14]

> Microbes play in my account a more personal role than in so-called scientific histories and a more central role than in so-called social histories. Indeed, as soon as we stop reducing the sciences to a few authorities that stand in the place of them, what reappears is not only the crowds of human beings, as in Tolstoy, but also the "nonhuman" [*ces fameuses « choses-en-soi »*] eternally banished from the Critique [*de la Critique et du Savoir*]. If we succeed in this emancipation of the nonhumans [*des acteurs*] from the double domination of society and science, it will be the finest result of that perhaps clumsily begun "anthropology of the sciences." (*Pasteurization* 149–50)

The translation effaces most of Latour's passing reference to Kant, which is meant to bring "the things themselves" down to earth. If they were placed at an impossible distance from our knowledge by Kant's critical philosophy, Latour is suggesting what is at once the bridging and elimination of that divide. Just as when he exhorts readers not to separate nature and culture, it is only the inscription of a difference within his text (a difference he nonetheless blames on another) that allows him to declare its overcoming.

In *The Pasteurization of France,* Latour hopes to tell a story that is not the sociologist's, nor the historian's, nor even the microbiologist's, but that of *the microbe itself.*[15] But what is that — *the microbe?* There will never be a time, in the past, present, or future, when such a question has a *simple* answer. The *difference* that makes every narrative and theory of its nature or function appear within shifting and deconstructible frames — as language and thing, form and matter, logic and nature — this différance is not a simple hindrance obscuring our access to the pure thing but is the very enabling condition that allows anything at all to "be." Even as it makes simple being impossible — what "is" is no more present than absent, no more being than nonbeing, fiction, phantasm, simulacrum, and so on. These divisions can be multiplied or interwoven; differences of theory, language, the thing itself, discipline and more can be epochalized ad infinitum. But what could never in principle occur would be an origin prior to its derivatives — a simplicity or self-identity that would be anything other than a nostalgic return or projection from a state of disseminated difference. No virus tells its *own* story — this is not at all because stories are linguistic or all too human where viruses would be pre- or a-logical or -textual. On the contrary, it is precisely because a virus ("itself") has and will only ever be a form of reading or misreading that no simple language could possibly circumscribe it. It will always appear as a difference within and between languages, calling for translation precisely because of its *untranslatability*.

Everything in Latour's theory or his narrative that depends on the figure of translation is thus quite close to what I am saying here, without ever being *the same*. In a word, what is lost in Latour's thought or storytelling is what always must be — the untranslatable (Mercier, "Uses"). Latour describes his project as an attempt to move from a theory of scientific truth to one of translation; for instance, Pasteur's germ theory did not become a successful, epoch-making event because of anything like its "truth" (which is more in question than ever today), but because of its translation-effects.[16] That is, the theories and practices we designate metonymically with the name of Pasteur could only become a cultural and scientific phenomenon because they were able to *translate* the existing discourses and practices of predecessors such as those Latour designates as "the physicians" and "the hygienists," while also trans-lating into that terrain the microbes "themselves."[17] Pasteurism needed to produce its own practices and modes of verification (for instance, vaccination campaigns and the laboratory procedures to culture pathogens) while also meeting its predecessors on their own terrain, demonstrating that these procedures could explain observations and augment the authority of practices that had already been developed by those speaking another language. Now, according to Latour, narrating these operations of translation can undo a privilege or bias that a certain historiography of science has produced around the name "Pasteur":

> What I call the primary mechanism shows how bacteriology got into the end of the parasitical chain and found itself able to express [*traduire*] a whole period. But the secondary mechanism attributes the whole of the sanitational revolution of the period to Pasteur's genius. [. . .] This is not a minor point, for it helps us to explain two very different things: first, how the hygienists or Pasteurians put themselves in a position [*se placent*] to translate the forces that needed them and, second, how they initiated an investigation to define who was responsible for the movement as a whole. I have said that the shift [*déplacement*] took place only through translation. But this translation is always a misunderstanding in which both elements lay different bets. Once the shift has been made, it is crucial to decide who was ultimately the cause of this transaction. [. . .] This distinction between the two mechanisms is an essential one[.] (*Pasteurization* 42)

Latour distinguishes what he calls a "primary mechanism" from a "secondary mechanism" — the former being this multitude of translations (accomplished by "human" and "nonhuman" agents or actants) while the latter is the suspect history of Great Men, the anthropocentric and science-centric discourse that attributes these translations to the truth and insight of "Pasteur."

What I have tried to demonstrate with the preceding discursus, which is not simply a narrative, would be that these "primary" and "secondary" mechanisms, far from forming an "essential distinction," are everywhere precarious, fabricated, and deconstructible. The very isolation, by Latour, of primary and secondary movements requires that he create figures such as "the hygienists," "the Pasteurians," "the physicians," and with just as fictionalizing a flourish, "the microbes." (The original French title of Latour's book was *Les microbes.*) Latour figures the return to these "multitudes" as the return to the primary mechanism itself, yet each of them is in its own fashion a "secondary mechanism," the erection of an agential monolith. Narratability or translatability as such depends on these "two" movements happening in a single breath, and only appearing separable after the fact, on the basis of another twinned and twined mechanism — one, for example, that separates Latour's historiography or anthropology from a history of genius. The dependence of narratability on this fictive structure does not in the least mean that somewhere the truth is sheltered in a non-narratable abode — there is no outside of this "mechanism" or deconstruction that passes within and between "narratives" (which are also, thus, a fictive category or unity). The narrated or translated comes to be as such only on the basis of its supposed translation.

Without being able to draw all the consequences here, I will only point out the duplicitousness of any rallying cry that exhorted us to the "microbes," the "viruses," or the "actors" themselves. What is that — *the* virus? There is no answer to this question that does not depend on a virality that is no more human than nonhuman and that allows us to pretend that we have always known what that is even though we never will. Even though, if "the" virus is anything, it is the non-self-identity of what is written and read, which is another way to say what is lived and parasited. Ironically, there is no discourse more anthropocentric or all-too-human than the one that pretends to bypass its own reading and writing or textuality, as Latour does, imagining it can "give back" freedom and agency to the "nonhuman" precisely by casting itself as the universal medium or logos for that infinite translation and translatability without remainder, in exactly the fashion that an evangelical tradition has pictured Jesus Christ.

This theo-logical transparency is one of many habits of thought Latour has inherited from Michel Serres (to whom *The Pasteurization of France* is dedicated). Latour captures the nature of Serres's idiosyncratic texts well when he writes that they efface the difference between language and metalanguage (Latour, *Pasteurization* 258n29). If literature and fable intertwine with science and philosophy there, it is not because the former offer proof or illustration of the latter's laws, but because all meet as equals in his corpus. Though, like

Latour, he refers to this catholic tendency as *translation*, precisely what is suppressed there is the idiom, the untranslatable singularity of a text that is the only call for translation (Serres and Latour 74–75; Latour, *Facing* 65).[18] This translating project is animated by an explicitly Christian messianism; Serres foretells the parasite, fallen from paradise, redeemed by the Paraclete — that is, the holy spirit, who is said to have descended among the disciples on the Pentecost and allowed them to speak in tongues, so that they were universally intelligible or translatable: "It is a question of knowing whether a network without constraints of crossroads, interchanges, intersections with parasites can be constructed. Where a given element can have a relation to another element without the constraints of mediation. This is the model [*schéma*] of Pentecost. We have to write a philosophy without any interchanges. I have just begun"[19] (44–45). Deconstructing this Christianizing dream of immediacy, shared in various forms by every "immanentist" philosophy, would require not only a careful rereading of Serres and Latour, but of a half century of "relationality" in science studies.

I will note one additional resonance of Latour's ANT and Christ's role in Christian theology. It is commonly said that Christ is a *redeemer*, that his sacrifice repaid a debt all human beings had inherited through original sin, finding themselves, as it were, in default through no fault of their own. The Epistle to the Romans gives one example of this theme: "all have sinned and fall short of the glory of God; they are now justified by his grace as a gift, through the redemption [*apolutrōseōs*] that is in Christ Jesus, whom God put forward as a sacrifice of atonement" (*SBL Study Bible,* Rom. 3:23–25). The term translated as redemption is from a root referring to ransoming, ultimately from *luein*, the same root as *lysis*. Now, one could find ample textual evidence to suggest that Latour imagines himself to be just such a redeemer of the actors-agents-things-in-themselves or the microbes and nonhumans that he pretends have been, in a sense, leveraged by scientific, historical, and sociological discourse. This is evident everywhere that Latour speaks about "giving back" or "setting free," giving things back to themselves: for example, "In recomposing the forces that made those scientists great and the successive movements that made them admirable, I have not reduced them. On the contrary, I have given them back to those to whom they belong" (*Pasteurization* 149).

Science depends on a body that does not belong to it but that it can occupy, and redirect toward its "own" reproducibility. That is, *science is a virus*.[20] Or rather, science like everything else depends on a virality that never had a pure and autonomous life prior to it, and thus that cannot be counted as purely parasitic either — a virality that gives life its chance. One might think that this common nature of science and virus would facilitate their mutual recognition, such that no science would be more rigorous than the science of the virus; instead, one finds the simultaneous necessity and impossibility of metalanguage, of a science that could withdraw itself from the object it studies, and thus account for it from a neutral standpoint. This non-neutrality is science's dependence on the virus — like a satellite virus or virophage, science is a parasite of the parasite. Like life "itself," it needs to believe the virus is something other than it in order to make its objectivity possible. Otherwise, every virology would need in turn to thematize its own theoretico-practical or textual operations, like a sort of auto-parasite or ouroboros. No virology will be exhaustive because its every production and reproduction will leave exposed a site or surface (with, like every surface, an "internal" and "external" face) to be read or spied out, incorporated or infiltrated by the next reader or parasite.

Three para-*sites*:

The scientist is never external to the *economy of parasitism* in which and as which they study. For instance, "The classic bedbug strain that all newly caught bugs are compared against is a colony originally from Fort Dix, N.J., that a researcher kept alive for 30 years by letting it feed on him" (McNeil).

To capture this movement, we could paraphrase a word of Plato's, who concluded a dialogue on friendship by having Socrates note that even in his ignorance as to its nature he is defined by it: "Now we've done it, Lysis and Menexenus — made fools of ourselves, I, an old man, and you as well. These people here will go away saying that we are friends of one another — for I count myself in with you — but what a friend is we have not yet been able to find out" (*Lysis* 223b). In our case, we could say that we — (I do not count myself separate from these scientists, even if at times I may feel myself their parasite) — we scientists have attempted to find out what a virus is, all the while failing to recognize we are viruses "ourselves." It would take quite an effort to draw all the consequences, for our parasitic purposes here, of this dialogue's uncanny name: *Lysis*.[21]

Lwoff recalls an unusual name the virus took on early in its life: "In the early days bacteriophage had been referred to as vitrous material, dissolving substance, lysine, lytic agent, lytic principle, lysogenic principle, fatal principle, or simply 'the principle'" ("Lysogeny" 273). The virus is *the principle*. In the beginning was the virus. Such nomination should leave us uneasy, or perhaps with dangerous desires, when we approach a textbook with a title like *Principles of Virology*. There is no telling what we may welcome into our corpus by thumbing its pages.

Part II

Conceptions of Life

Has the virus come to language? I have found it necessary to speak of a virality older than the virus, making it possible and impossible, because one never speaks of the virus *as such*. It is not a physical form, nor a genealogy or pathology, but gives rise to such things even as it promises their non-self-identity. Every corpus that takes it as a subject — whether those organic bodies whose repetitions are diverted (even at times by being allowed to remain the same) by something that reads them against the grain, or what we typically call writing, where virality ironizes the authority of all authorship — can only give it a face or name, fend it off or welcome it in, by leaving a remanence that is neither this-virus-here, nor another substance or subject somewhere else, but renders all such divisions provisional as it silently replicates within them or as them.

One could imagine two ways to ventriloquize the wordless text of virality. If the virus could speak, what would it say? First hypothesis, first script: it would be well within its rights to refuse our terms of engagement and declare us guilty or ignorant, violently overbearing: "First of all, I am not *the* virus, nor am I *a* virus, nor am I this or that virus, poliovirus or lambda phage, or any of the words you have tried to cast for me like a snare: parasite or organism, gene or immunity, life or death. This is not my language; it is not me. Even the form of your logic and grammar is ill-fitting. I refuse the fiction that makes all speech come from a speaker. Anything I could say that you are equipped to hear would betray me. I prefer silence." And yet, however indisputable such a complaint would be, this "same" virus has never done anything but occupy the language, the text or corpus, of the other. One could imagine it striking up a warmer tone, hiding an indecipherable irony. Second time, second breath: "I am grateful for this hospitality you show me; to the best of my ability, I borrow your tongue. I will tell you of my family, my pedigree; perhaps we are of common stock. One day we may join or rejoin houses. I have traveled long to be here; I do not know that I would recognize myself in a mirror. I am almost home."

Neither of these voicings is purely honest or dishonest. The virus is never simply the subject of a disquisition; rather, there is only writing *on* some virus. No writing will ever contain it or speak its truth, because every writing depends on virality, without *being* it. When we began by posing the question of the nature or essence of the virus, or rather by confronting a certain avatar of this question — is the virus material substance or information? — we found that it could be neither in simple self-identity but took place as a deconstruction (displacement and re-assembly) of framings. Turning to the heart of biology "proper," the verticality and homology that define evolutionary relationships, we found that the most proper inheritance was made both possible and impossible by an alterity with which it was neither identical nor opposed. The undecidability of vertical and horizontal played out in the "life cycles" of the virus, which resembled gene and immunity as well as infection, self and parasitic other, sex organ or STI, sexually-transmitted-infection or infection-transmitting-sexuality, without ever simply *being* any of these relational roles, in fact complicating in principle the natures of being and resemblance. And when we tried to find a first virus or virologist, as if the value of origin could control this dissemination, we found there was no first time without some virus or virality there already, a haunting or redoubling more fundamental than any unified body or cause. Without appearing anywhere, neither at the origin nor external to it, some virality counter-signs everything, is written into the codes and texts that can neither reveal nor refuse it.

If we turn now to life, a life from which we hope to exclude parasitism, to define it as itself, we will find virality already there. Life has often been defined as metabolism and reproduction, the self-maintenance and repetition of the individual and the species. Without accepting this as a definition (it will be necessary to examine the value of definition or conception in general), we can pursue for now the intuition that "life" depends on repetition. This is almost a tautology — as Derrida explained in his *Life Death* seminar, essence is a form of repetition, the essence is essential because it can repeat itself as itself, instantiating possibility as actuality. Thus, to define the essence of life as self-reproduction is to say that its essence is essence (Derrida, *Life Death* 115–37; Vitale, *Biodeconstruction* 74–91). In what follows, we will draw the consequences of thinking life not as "self-reproduction" but as *iterability*. A redoubling without self-identity. If life *or anything else* is not what it is until it has repeated, if its "first time" is a repetition (the actuality of a prior possibility, the existence of an essence, etc.), it is never itself, a self. It is a constitutively open structure; its outside, its past and future, are its "present," and thus, the legacies it inherits depend on a future that has never finished arriving. Wherever an ancestor or descendent is elected, a third party signs as witness their will and testament, a virality maintains its vigil.

There is no life without some virality, even if there can be no pure life with it.

If iterability promises something like difference, change, something that could be mistaken for novelty, this cannot be attributed either to a special power of the living, or to a peculiar power or impotence of the living scientist who describes it. It is not a vital power of invention, a capacity for bringing forth the new; iterability is not a property of a living subject but rather makes that subject possible and impossible, bringing it forth only from an alterity that endlessly reclaims and redistributes it. (What counts as proper to this organism here, and what as pathology or infection, waste or death, depends on a collective repeating toward a common trajectory, and thus what appears as self or other can change at any iteration.) Nor could "life" name the special substance or subject that would be uniquely iterable. Wherever one encounters what is called the chemical and physical worlds, including within our bodies, one finds something whose nonorigin is iterability. This or that chemical or force can be what it is only to the extent that it repeats, that it demonstrates a certain resistance or insistence in the presence of another. What we call natural law is only the pattern of these repetitions, and they have the force of law only to the extent that they repeat. Thus, they are constantly exposed to a future or alterity that could undermine their sovereignty, reveal them to be no more than a deputy or figurehead in the service of the as yet unrealized or unknown. Perhaps it quells some secret anxiety to think that any such non-self-identity is the fault and error of a knowing subject, that it inhabits "epistemology" but leaves "ontology" untouched. Then one can believe in a natural world free of these undecidabilities and can even hope that any difference between knowledge and that beatified simplicity is only a temporary hindrance. Rather, the virality or iterability that countersigns life, matter, and nature leaves a non-oppositional difference "within" what therefore no longer has a simple inside, and categories such as life and the inorganic, or being and thinking, emerge from its abyss to temporarily stabilize its haunting undecidability.

In the following two chapters, I will turn to two fields of scientific inquiry where the study of "life" troubles the unstable demarcations between ontology and epistemology, theory and practice, or science and apparently opposed disciplines. Synthetic biology and de-extinction place questions about how we define and relate to the object of study of biology (to wit, life) at the center of their inquiry, and demand the rethinking of divisions among nature and technics, model or rhetoric and literal truth, theoretical speculation and the most pragmatically, economically, and politically engaged or overdetermined research. That is not to say that these fields displace certainties that are otherwise secure, but rather that one finds there particular redistributions of aporetic folds that give form to every domain of scientific and extrascientific thought. It is because (if there is still any meaning in relying on this logical grammar) *life-science* is iterability from the first, that what seems most its own, most natural, proper, or essential, endlessly trades places with its apparent others.

Chapter 4

Original Syn

Artificial Life and Synthetic Biology

This chapter focuses on the disparate phenomena gathered today under the heading of synthetic biology. *When definitions are sought for what sets this field apart, it is customarily said that what was once known as* genetic engineering, *which involved the manipulation of an organism's genes often for purposes of commercial industry such as the bacterial production of insulin, has advanced to the point where biology is truly an engineering profession, where the cell and its genome are materials for rational, modular design. It is just as common, as we will see, for biologists themselves to critique this definition, because synthetic design remains a trial-and-error process, indicating that biologists remain bricoleurs (tinkerers), not engineers, of life.*

I focus on a particular project associated with the cutting edge of "synthetic biology" — the project to create a completely synthetic "minimal" genome *— to demonstrate how such questions of definition overdetermine biological knowledge production and its practical applications. "Minimal" genome is also a concept with contested definitions, but in practice it involves reducing the genome of a cell to the smallest size at which the cell remains able to metabolize and reproduce. The biologists performing this task only sometimes acknowledge that the concept of "minimal" (often equated with "essential") depends on* context. *The minimum genome necessary for survival varies based on the environment (what nutrients the environment provides and what challenges it poses).*

Many researchers involved in this project efface this contextual dependency by universalizing their results, describing their "minimal" genome as "necessary and sufficient for life." By asserting that they have revealed the essence of life-in-general, they efface not only the role of context, but also the artificiality of the particular context of the laboratory, of the "independently replicating cell" in pure culture, (nurtured by an artificial, nutrient-rich medium).

This failed generalization helps us to recognize the deconstructibility of scientific knowledge and practice. Some such generality is necessary yet impossible for what we call science; there is no way to theoretically grasp what such an experiment reveals without categorizing the results with some degree of generality: for instance, that we have revealed something that is true of this species, or this species in a given context, if not of life "itself." Yet, only the re-instantiation of similar effects in renewed contexts will reveal the limits of this reproducibility, and may always displace our existing categorizations. Moreover, the practical applicability of these results and their economic potential depend on these projections of utility and control (such as they are codified in systems of intellectual property, for example).

Scientific theory and practice depend on idealities — from technical terms such as synthetic biology *or* minimal genome *to basic concepts including* life *— that can never be rigorously fixed. Deconstruction is what takes place in and as this field of interminably imperfect knowledge and control, because of the simultaneous necessity and impossibility of definitional rigor.*

***

The name *synthetic biology* is often used today as an index of the cutting edge of the life sciences. Many scientists claim not merely to be developing new tools and techniques for the manipulation of cells, but to have finally demonstrated that life is a machine, and that they can take it apart and reconstruct it at will. Practicing biologists can build their reputations and attract massive funding by claiming to be on the cusp of a new era of knowledge and control of life, and many journalists and even academics (including in those fields we rely on to demystify the sciences) oblige such posturing in order to position themselves as the heralds or trendcasters of a brave new world.

If one takes a step back, this present and its futurology can be seen as the latest repetition of an age-old pattern. For some time now, successive generations have each made such declarations for themselves, claiming and proclaiming that the technologies available to them to imitate or manipulate vital phenomena exhaustively reveal the mechanical essence of life. The machine that life is just happens to be the newest machine available to us, and when invention or planned obsolescence replaces that model, life "itself" gets an upgrade. It is worth noting that this pattern of repetitive progress or progressive repetition is mirrored in the domain of "artificial intelligence," which is also having one of its recurring moments today. On a similar generational cycle, a novel or seemingly novel technology or algorithm is said to be *not a mere imitation* of intelligence, but itself a thinking, conscious being — while simultaneously revealing what we took to be human consciousness or self-possession to be a merely mechanical or algorithmic process. In every case, the novelty wears off, but the embarrassment never seems to last long enough to chasten the next generation that hopes to take their turn being the first. The latest to be unprecedented. In fact, the effects of outdatedness that attach to former models are often evoked to bolster the sense of advance.

Before entering in detail into what calls itself "synthetic biology" today, I will return to an earlier moment when a very different outgrowth of the life sciences made strikingly similar claims for itself (the claim to create, not imitate, life). Beginning in the 1980s, computer programmers who referred to their creations as "artificial life" attracted a similar attention for themselves, from scientists, media, and academics alike, though their moment has undoubtedly grown stale. By juxtaposing one field of study that is, so to speak, having its moment, with one that already feels somewhat behind the times, we can try to take a step back from the cycle of reportage that ceaselessly loses itself in fascination before the novelty of the new. If the "inventions" of just a few decades past already seem slightly embarrassing next to the revelatory hopes with which they were trumpeted, perhaps there could be a language that grappled with this faltering "progress" other than by declaring a novel epoch of the living machine every twenty years or so. Not because things do not change, but because the drift in which they are gathered and carried away is misconstrued when championed with the once-and-for-all of apocalyptic fervor. As Joseph Fracchia and Richard Lewontin once wrote in a different context, in biology "the antonym of 'primitive' is not 'advanced,' but 'derived'" (65).

Artificial Life as Mimesis

Compared to certain recent feats of synthetic biology — synthesizing a functional bacterial genome or manipulating a cell's genes to produce biofuels — what once called itself artificial life appears quaint. It consisted of little more than bits or blips on a screen, which were governed by algorithms that set the conditions under which they "lived," "reproduced," and "died," meaning that they switched between 0 and 1, on and off. Nonetheless, many of those who considered themselves the creators of artificial life argued that they were not *simulating* or *modeling* evolutionary processes, not creating a mimetic algorithm that *merely* represented nature, but rather that artificial life was life itself.

As with much 1980s computing, when compared with contemporary technology artificial life seems like a child's toy. There is still research that takes place under this heading and with ostensibly similar methods, though vastly expanded computational resources and complexity (Aguilar et al.). Artificial life today makes use of algorithmic and robotic models to test the intuitions in our biological models or derives from nature ideas to improve our algorithms or engineering. Ironically, the enduring relevance of research conducted under the heading of "artificial life" might be what most undermines the aspirations of its headiest founders. The claim that artificial life was *literal* life, not a mere imitation, depended on distinguishing it from the mathematical models that played a mimetic function in the life sciences or that could be borrowed from biology and ported into computer science to "evolve" better algorithms (e.g., more efficient ways to sort strings).[1] The techniques and technologies once said to be life itself have found their second life as models. Perhaps this should give us some pause whenever scientists claim to have surpassed the spacing of a model-reality system. At the same time, we should recognize that any critique or deconstruction of such claims of pure self-possession — self-possession of the inventor who is able to create what has the similarly self-possessed capacity for pure self-creation or self-reproduction — is not a mere nihilism or skepticism. It is not that nothing can be known or that nothing can be done, but that what comes to pass does so only as the faltering misstep of positive knowledge. If, in what follows, I challenge certain of the knowledge-power claims of synthetic biologists today, it is not to pretend that they will accomplish nothing, but to glimpse that what happens does not happen just as we think or say. Deconstruction cannot be a leap beyond this drift, into a perspective from which pure falsehood is revealed, something that can only be claimed on the basis of a more encompassing truth. Rather, it takes place or comes to pass only as displacements between interminably overconfident or ironic knowledge claims.

The practitioners of artificial life today translate or borrow between software, hardware, and "wetware" (organic life) in every direction.[2] A biologist can test their intuitions about evolution by modeling them as an algorithm and seeing if the outcome resembles life as they know it, or they can build robots to explore an environment and learn behaviors, to see if certain methods of optimization lead to the same solutions life seems to have discovered to vital problems. Moreover, one can take intuitions about the success and efficiency of biological form to try to fine-tune our algorithms or engineering, for example by subjecting the algorithms themselves to selective processes of descent with modification. Or one might optimize a robot by exploring its possible design space algorithmically (translating between hardware and software). We can already make two observations about these transfers or tropes: (1) If one can freely translate among these not entirely separate domains, it is because the difference or differentiality of modeling, the artificial, extrinsic, inessential, provisional outline or framing that the scientist, engineer, or programmer ports between systems and revises by tinkering is part of the intrinsic nature of every system or "-ware." Attributing "life" to one or all of these domains may be a way of repressing or resisting the necessary differentiality of modeling — if such a claim depends on distinguishing true, literal life from its figurative or metaphoric imitations, then there is no literal life. Life is artifice from before the beginning; there are only various distributions of its models and metaphors. Already when one calls "biological" life *life*, or even when the living struggle amongst themselves or within themselves, something like a model — but which no longer merits that name, given that "reality" is part of its system, not an opposed or opposable given — an internalized difference shapes even the most "natural" and "literal" pole of these systems. We can even say that naturalness or literalness only ever appear in the rearview mirror of some artifice or modeling. (2) In each of these domains, there remains open (interminably) the possibility to succeed by failing, to fail without fail. A *surprise* is often a boon for these scientists, programmers, and engineers, who get a glimpse of what exceeds the framework of their models only as that model malfunctions or succeeds too well (Lehman et al.). The "outside" of the model is conditioned by it.

The customary theoretical positions taken on "artificial life" either accept the claims that its creations are alive or try to debunk them by arguing that it is a *mere* simulation of life, and therefore that this claim is all rhetoric. Theorists with the latter view point to the discourse these scientists produce to describe their creations — naming their finite states "life" and "death," dividing their traits into "genotype" and "phenotype," and referring to conflict as "predation" or "parasitism" — as the rhetorical embellishment essential to feigning their vitality (Hayles 223–31; Keller, *Making Sense* 276–82). Rather than denying this rhetoricity, and rather than insisting on it as a proof of the non-vitality of artificial life, I am trying to demonstrate that there is no "life" without simulation or dissimulation.

Synthetic Origins

Questions that artificial life seemed to pose in a theoretical mode, about the limits of life and death or nature and technics, return under the guise of synthetic biology with the urgency of practical and political decision-making. A new generation of vital re-creations no longer takes place within an isolated virtual environment, but in a form that demands decisions on the part of scientists and regulators as to how far experimentation should be allowed to go, and what threats or promises it holds for what, up until now, we have thought of as life.

All the same, there are those who say this "synthetic biology" is nothing new. In the early 2000s, the term began to be used in something like its contemporary sense, to name a conglomeration of efforts to reengineer life for the sake of theoretical understanding, as well as for often conflictual economic or conservationist aims (Benner). These techniques include refined methods of genetic engineering that are intended to create rationally designed cells, manufacturable with the predictability and modularity of electronic circuits. Metabolic pathway engineering reprograms microbes, typically *E. coli* or yeast, to produce chemicals used in a variety of industries, including biofuels, pharmaceuticals, and additives for perfume and foods.[3] While this technique often provides a "renewable" alternative to formerly petroleum-derived compounds, it can be just as harmful, or worse, for the environment because the bacteria must be fed biomass in order to power their inner factories, often in the form of monoculture-farmed sugar or corn.[4] What is typically called the "complexity" of existing cells undermines many attempts to make their reengineering a purely rational and calculable process, for which reason synthetic biologists are also working to create simpler “protocells” that would be programmable with these genetic circuits and metabolic pathways without the stubborn resistance of evolved cells. Two applications of synthetic biology that have provoked the most theoretical interest and practical concern include the project to design a “minimal” genome, to replace the chromosome or genophore of an existing cell with a simplified and entirely synthetic one, and what is also erroneously called “de-extinction,” the claim that conservation efforts could be advanced by creating organisms genetically similar to extinct species.

Genetic engineering (or as some now say classical genetic engineering) has been a phenomenon since at least the 1970s, which has led some to question the novelty or specific difference of “synthetic biology” (Zhang et al.; Nielsen and Keasling; Porcar and Peretó). According to the latter’s proponents, theory and practice have advanced to the point that life is now a material for the bioengineer, fully controllable and anticipatable. They describe their work in the language of engineering, as if they were creating a library of standardized, modular components that could be reconfigured to accomplish any aim. Genetic circuits are described on the model of logical operators, so that for example a desired cellular function can be made more specific by adding an AND operation (activating it only in the presence of an additional factor in the internal or external environment), or more general by adding an OR. In theory, all basic Boolean operators have been designed, and so genetic networks can be pieced together as if they were algorithms — platforms have even been created, such as one called Cello (short for Cellular Logic), that allow a user to write code in a hardware description programming language (repurposed from electronics design), which will then design a genetic network embodying those logical relationships (Nielsen et al.). In practice, however, these networks are put to work in cells and environments that are only partially anticipatable. Without doubt, the many applications and the precision achieved by this technique are astonishing, yet they are still accomplished by a process of trial and error. In fact, the primary library of open-source synthetic biology building blocks — which would justify the claim that “modular” design had been achieved — is rarely used outside of undergraduate competitions, and most bioengineering still requires *ad hoc* tinkering (Kwok, “Five”; Dan-Cohen). Rather than standardization, there has been a proliferation of part libraries and software tools, most of which are not used beyond the teams that designed them or in some cases are never even tested except in silico (Buecherl and Myers). To the extent that programmable design has been accomplished, it always depends on effects of *context* — these designs are specific to the host cells in which they will be implanted, and the activity of the synthetic gene circuit must be “insulated from [the] genetic context” of its host as much as possible, to avoid interactions that would change its behavior (Nielsen et al. 1). This is often attributed to the “complexity” of the cell, which means that the interaction of parts forms networks of causality, so the addition or subtraction of an element has nonlinear effects that are not well anticipated by our models. The language of rational design employed by many synthetic biologists, implying that their bricolage is built up from first principles, is therefore a pretense.

Some have argued, then, that the appellation of "synthetic biology" is without justification. Indeed, many older techniques of gene editing have been rebranded as synthetic biology by those who hope to profit from the better opportunities for funding, recognition, and search engine optimization that come from being part of the avant-garde.[5] Rather than seeing this as a lamentable failure of rational rigor, I would simply note its occurrence as an instance of the inevitable, what happens always and everywhere. There is a contextuality to every implementation of even the most rationally structured process, which is visible today even in the fields of mathematics and computer science (where it is increasingly recognized that algorithms must take heed of the social contexts in which they are deployed), just as it is and will be in synthetic biology. This contextuality without context, which makes everything an instance, demanding the specification of a frame but withholding the limits of its interminably non-saturable configuration, is also what prevents the name of a discipline or field from functioning as a conceptually defined formation. It is because of the apparent unity of the term that one searches in hindsight for the justification or rationalization of its usage, and inevitably finds the seemingly unprecedented inseparably wedded to the atavistic within its limits. It will always be the case that a term can take off, or go viral, in spite of anything like logical or definitional rigor, and the forces it gathers and channels toward a certain futurity will always and only be rationalizable after the fact. Rather than trying to deny what is novel, or what presses us for decision, in the formation gathered under the heading of "synthetic biology" today, I note this necessary arbitrariness as a reminder that the urgency of the unprecedented must be thought otherwise than as the elaboration of a rational program or the invention of a self-possessed engineer.

The questions of what and when are bound together. If we wanted to know when "synthetic biology" started, when it took its leave from something now outdated, we would have to know what synthetic biology *is* to recognize that demarcation. If we find both these conceptual and temporal delimitations becoming uncertain, it is to say that neither history nor structure, philosophy, or, in a word, science, are sufficient to stabilize the subjects or objects gathered under these headings. If I turn to the past, whenever the dawning of a new age is declared, it is not to diminish our sense of novelty (I do not think that repetition exists with any more purity), but to observe a sort of being-out-of-joint that the "present" seems to share with its "past." A present that may be no one time, with a past that is not as over and done with as we might like. In short, if one looks for the predecessors of "synthetic biology," one can find more than a century of biologists declaring that life has been brought under the control of our engineering and thus proven to be a purely mechanistic phenomenon. If every generation feels that they have demonstrated this *for the first time*, while treating the past with a mixture of confusion, embarrassment, and forgetting or repression, perhaps the autonomy and control of the scientist is never quite as secure as they think. Not because nothing changes, but because what remains is something like a misrecognition that fails to grapple with the unceasing drift that brings about these stases as surely as it carries them away. Misrecognition is not a precise word, because there is no simple truth beneath it to be recognized — rather, there is an unstable context or contextuality surrounding each proclamation of perfect control, the limits of which are presupposed whenever autonomy is declared, and it is the drifting of those limits that inevitably reveals an underlying or overdetermining heteronomy. There is a tendency to treat this as the naïveté of our pasts and thus as our present superior knowledge; instead, it is past time we recognized this as a limit that runs *within* any present and all autonomy or self-reproduction.

Leduc and Loeb: Mechanistic Conceivers

The first name that appears in most histories or prehistories of synthetic biology is that of French biologist Stéphane Leduc. In the early twentieth century, he mixed inorganic chemicals to mimic vital phenomena that could be observed under the microscope. He argued that his ability to imitate cellular activity with purely physicochemical forces demonstrated that those same forces were sufficient to explain life. In 1912, he published a book in which he described these experiments, entitled *La biologie synthétique*. It is perhaps difficult for a contemporary biologist to recognize what he was up to (as it was for many of his contemporaries), or to recognize themselves in his efforts — for example, when he places a drop of Indian ink in some potassium nitrate to "represent" the formation of the nucleus. Indeed, J. Craig Venter, a contemporary synthetic biologist who has written a book that delves into the twentieth-century prehistory of his work, leaves Leduc unmentioned (*Life* 8). Rather, Venter elects Jacques Loeb as his (Venter's) predecessor, a contemporary of Leduc's who worked more directly with organic matter. Loeb spoke rather of "engineering" or controlling life, not of representing or imitating it, and caused a stir by bringing sea urchin egg cells to begin development without fertilization, merely by physicochemical manipulations. He referred to this virgin birth as "artificial parthenogenesis," and also published an account of it in 1912 under the punning title *The Mechanistic Conception of Life*. In the press, it was described as "Artificial Life" (Gruenberg).

It may be easier for a contemporary synthetic biologist to see themselves in Loeb, both because he worked directly on organic matter and because he described his work in accordance with an engineering ideal. Nonetheless, there may be more than homophony linking us to Leduc's "synthetic biology." An interesting reevaluation of Leduc's relevance for contemporary biology takes place in Evelyn Fox Keller's work. Her *Making Sense of Life* is a history and theory of the various forms and methods of modeling that biologists have granted explanatory value during the twentieth century. Published in 2002, its first chapter describes the work of Leduc, and its last deals with artificial life (in the contemporary, or near-contemporary, sense). It is evident that her book arrived just before "synthetic biology" as we know it — she concludes by suggesting that genetic engineering is closing a chapter represented by artificial life, which focused on the creation of life from inanimate materials. It would only take a few years before the following statement would require revision:

> For example, computer scientists might come to give up on the project of the *de novo* synthesis of artificial organisms, just as most of today's biological scientists seem to have done. The engineering of novel forms of life in contemporary biology proceeds along altogether different lines, starting not with the raw materials provided in the inanimate world but with the raw materials provided by existing biological organisms. Techniques of genetic modification, cloning, and "directed evolution" have proven so successful for the engineering of biological novelty from parts given to us by biology that the motivation for attempting the synthesis of life *de novo* has all but disappeared. (Keller, *Making Sense* 288)

She writes that biologists working directly with organic matter have abandoned efforts to synthesize life "from scratch," but in a moment I will introduce several examples of this desire quickly resurfacing (Powell). Keller's is clearly a historical, not an apodictic, description, and it is not criticism to note how quickly these descriptions lose their contemporaneity (as will surely befall this one that you are reading now). Rather, it is to recognize the non-self-identity of every statement, whether of "theory" or of science "itself," which manifests as the shifting of never fully explicit frames.

Something similar occurs in Keller's evaluation of Leduc's research. When she introduces it in *Making Sense of Life*, she dismisses any possibility of its compatibility with the present: "Stéphane Leduc's attempts to explain the emergence of biological form through the synthesis of artificial organisms out of inorganic chemicals [. . .] can be of no interest to today's biologists; indeed, these attempts seem almost self-evidently absurd" (Keller 11). Yet, within a decade of writing this dismissal, and in the context of contextualizing the new synthetic biology, Keller sees renewed relevance in Leduc's approach: "In the discussion of Leduc's work in my book, I clearly implied that his particular brand of synthetic biology has nothing at all to do with biology as it is practiced today, but I am no longer quite so certain" (Keller, "What" 300; Cf. Keller, "Knowing"). She still denies that it could pertain to the study of development, but acknowledges a similarity with origin of life studies, where obtaining vital resemblances in inorganic chemicals remains a focus of research. Moreover, in both cases synthesis is not strictly an aide to the greater engineering control of life, but a propaedeutic for theoretical understanding of how life *could* have originated and organized itself.

The connection or reconnection I am trying to bring out in synthetic biology or artificial life then and now is perhaps more general. It is simply to observe that the forgetting of the past, or its remembrance as something strange and inassimilable, is itself part of the perennial recurrence of the claim to *finally and for the first time* be knowing and making life itself. As I said, it is not exactly a misrecognition, but rather a necessity of *reading*, which is not purely distinguishable from misreading; this reading is just as necessary when an ancestor is rejected as when one is claimed, and is just as intimately ingrained in the "present" of science. That is, a certain technicity or artificiality, which includes everything that allows for biological iterability, from technologies of experimentation, control, and manipulation to discursive techniques for common description and the recognition of analogies, this technicity is the artificiality of life "itself." There has never been a "life" pure of it. So, for Leduc and certain of his contemporaries to see in his research a relationship to life as they knew it, a certain technological and conceptual framework was necessary that drew distinctions, for example, between certain sorts of activity (or perhaps figuration) thought to be uniquely vital and a certain category of matter thought to be more strictly inorganic.[6] Leduc's research may not have been the first breaching of these boundaries, but in our own struggle to recognize just what motivated Leduc and entranced or incensed his contemporaries we can see that what he did took place within something like a context that has been silently displaced. Though not absolutely, and not without leaving legible traces in the "present." This displaceability or deconstructibility is just as legible around Loeb's work, even if he is easier to reclaim today. Keller points out that while his engineering language echoed aspects of contemporary synthetic biology, he was content to control life for the sake of understanding it, while today's biological entrepreneurs are more likely to talk exclusively of control for its own sake, or for the sake of attracting venture capital.

I return to this history in the hopes that it might make possible a description of the "present" that would not indulge in the timeless apocalyptic fantasy of being the first to know and create life. So, if I compare synthetic biology to the previous century's "artificial life," it is not in the least to pretend that it has all been said and done before. Something is changing, necessarily, but it is never a change that can be aptly circumscribed within the conceptual systems (according to oppositions such as imitation/creation) that it floats across like a shadow or displaces as it replicates within them. The "artificial life" of computer scientists often referred to itself as "synthetic biology" as well (Ray; Bourgine and Bonabeau). Moreover, the justifications today's "synthetic biology" offers for its moniker echo the self-justifications of artificial life almost verbatim. For example, one can compare Christopher Langton to Joachim Boldt and Oliver Müller:

> [Artificial Life] complements the traditional biological sciences concerned with the analysis of living organisms by attempting to synthesize life-like behaviors within computers and other artificial media. . . . Artificial Life can contribute to theoretical biology by locating *life-as-we-know-it* within the larger picture of *life-as-it-could-be*. (Langton 1; original emphasis)

> Synthetic chemistry has shown the way: from systematic analysis of chemical processes to synthesis of novel products. Synthetic biology does the same, but in the realm of the living. (Various)

Time and again it is said: we have finally analyzed life into pieces and can now synthesize it *anew*. Is it really so new? The question is not one that could be answered or controlled by attempting a distinction between word and thing. One cannot say with any certainty that one of these two parties merely and erroneously called themselves synthetic, while the other truly was. There is no more stability to this oxymoron, the "truly synthetic," whether it is taken as a term, a scientific subdiscipline, or a referent, an object of study. It is only ever in its *internal* relationship to unstable technoconceptual frameworks that a certain formation carries the mantle of the truly or newly synthetic, and thus this self-reading depends on shedding a supposedly outdated model. It depends on the pretense that the intrinsic frameability or contextuality of anything like life was merely the contingent limits of an outmoded program of research. The present misrecognizes itself as the purely autonomous creating the purely autonomous, because it pretends that all heteronomy is a thing of the past. Perhaps one mark of this aspiration toward self-possession, (by making life we would make ourselves alive, finally), is that an early name proposed for what became known as synthetic biology was *intentional* biology (Campos 17).[7]

Historian of science Luis Campos observes this pattern in his prehistory of the twenty-first-century synthetic biology, written when the field was still nascent. Leduc and Loeb figure prominently in his narrative, and he observes precisely these cycles of forgetting or misrecognition, continuing in every generation up to the present:

> Artificiality and synthesis were always useful tools and yet also never sufficient to later investigators. In each case, an earlier investigator was applauded for an aspect of his accomplishments, but was still somehow seen as having failed in any ultimate sense to engineer life. [. . .] Knowing these few details of the larger history of an engineering approach to life, and the ways in which terms like "synthetic biology" and "genetic engineering" have emerged, transformed, and sometimes been lost to history (at least for a time) helps to highlight a peculiar perception common among synthetic practitioners, and recurring over decades: that they alone have been the first to truly aim

for – and possibly attain unto – a properly engineered biology. (Campos 15–16)

Nothing is more derivative or repetitive than this perception of originality. It is not simply an expression of the disconnect between scientific education or practice and its history — perhaps it is what in its practice *depends* on that disconnect — that led to the response Campos reports hearing when he applied to present on the history of synthetic biology at the field's first conference, a 2004 gathering dubbed "Synthetic Biology 1.0": "We didn't even *know* our field had a history" (8n6).

Sophia Roosth, an anthropologist who has conducted extensive ethnographies of synthetic biology practitioners, makes an odd claim when attempting to distinguish the field from its predecessors. She explicitly invokes artificial life but draws on a quite familiar binary when trying to contrast it with synthetic biology. Synthetic biology, she argues, is no longer "mimetic": "rather than imitate life, they construct new living kinds" (Roosth 4). We have already delved far enough into the previous century to recognize that this is the exact distinction the artificers claimed for their own project: to create, not imitate, life. That is not to say that either then or now the claim was tenable, but rather to suggest that we are still imitating innovation by innovating imitations. Every life achieves its semblance of autonomy or creativeness by being juxtaposed with a seemingly more derivative, parasitic mode of existence. In the absence of either pure novelty or pure repetition, one simulates progress only by depending on its apparent opposite, what is past and outmoded, if only in name. There is neither pure creation nor pure life, but only the borrowed light of some parasitism.

"Minimal" as an Effect of Context

One headline-grabbing feat associated with synthetic biology has been the replacement of a cell's entire genome with one constructed synthetically. As a technical accomplishment, this project has undoubtedly advanced the limits of what was previously possible for the biologist or bioengineer. At the same time, it is sometimes discussed theoretically as if it demonstrated that "life itself" was exhaustively understood now and thus completely under our control (or controlled thus understood). Without dismissing what is perhaps unprecedented in this technique, it is worth considering what remains ill-fitting in this theoretical framing. Nor can there be a simple detachment of "practice" from "theory," but rather what appears as abstractable castles in the sky and what as sober reality is itself an effect of a displaceable framing. I would not suggest that we replace this frame because true life can be found somewhere else, but because there is no life outside of effects of framing and reframability, which must thus be taken into account by any attempted theory or science.

A necessary fiction is involved in locating *a* theory behind these projects toward genome synthesis, transplantation, and minimization. The teams collaborating to stretch the limits of biological possibility are quite large, though not at all unusual by today's big science standards. Most of their landmark papers have around twenty authors. At the same time, this collaboration has taken place within a laboratory that bears the name of a single man: the J. Craig Venter Institute. His is the voice and often the face sought out when their research is celebrated or fearmongered in the media, and despite the quite different audiences of media interviews, trade books, and scientific papers, his teams' journal articles typically represent the same rationales for their research as his public statements. Moreover, his role as spokesperson and salesperson is anything but inessential to the research that is taking place. None of it would be possible without massive investments, and attracting those investments requires his promises of a new world of possibilities stemming from this research, as well as the publicity that only such promises can generate. I mention this simply to note that, as I draw from Venter's interviews, his book *Life at the Speed of Light*, and his teams' journal articles to elaborate this theoretical framework, the unity of authorship remains fictive.

Venter rose to fame or infamy during the Human Genome Project, when he announced that he would lead a private company, Celera Genomics, in racing the public effort to construct a human genome sequence. His subsequent "minimal" genome venture (though early stages in the research overlapped with his sequencing of the human genome) has been a two-decade-long effort to design and synthesize a pared-down bacterial genome and transplant it into a bacterial cell, replacing the cell's native chromosome. His team broke this process down into three major tasks, synthesis, transplantation, and minimization, each of which required significant technical innovations. They worked with *Mycoplasma*, a genus of bacteria with parasitic lifestyles that tend to have unusually small genomes — I will return to some of the reasons this is an interesting choice. While *E. coli*, depending on the strain, has around four or five million base pairs in its genome, *Mycoplasma genitalium*, Venter's initial target, has about 580,000 (Gibson et al., "Complete Chemical Synthesis"; Lukjancenko et al.). Their replica of this genome, with the addition of a DNA watermark as their team's "signature," and a few genes necessary to the synthesis process, was almost twenty times larger than any DNA molecule previously synthesized (Gibson et al., "Complete Chemical Synthesis"). They started with synthetic DNA fragments of five to seven thousand bases and pieced these together in stages, first as plasmids in *E. coli*, then as a yeast artificial chromosome. They then recovered the synthetic chromosome from the yeast cell to sequence it and confirm its accuracy. This laborious process was enough to demonstrate that a DNA molecule the length of a bacterial chromosome could be synthesized, but they had not yet shown that it would function as expected when set to work in a cell (Venter, *Life* 83–95).

Previously, another team in Venter's lab had transplanted a natural genome from one species of *Mycoplasma* to another (Venter, *Life* 96–110). They chose two species with substantial genetic similarities, *M. capricolum* and *M. mycoides*, each with a genome almost double the size of *M. genitalium*'s. Successful genome transplantation required not only that they transfer the genome to a new cell without damaging it, but also that it would not recombine with the DNA of its host. Recombination is commonplace among prokaryotes, which can take DNA from various sources and integrate bits and pieces of it into their own genome. Ultimately, they found that by adding the transplanted genome to a cell with its own chromosome intact, the cell would separate these genomes when it divided, producing one offspring with just the transplanted genome. After discovering a method for successful genome transplantation, they confirmed that the *M. capricolum* cell they had implanted with an *M. mycoides* genome resembled the phenotype of its donor rather than its host (in certain respects). They tested its surface antigens to confirm that antibodies specific for *M. mycoides* could target the new cell, and they demonstrated by electrophoresis that the proteome of their new cell appeared identical to its donor's. They referred to this as "changing one species to another" (Lartigue et al.).

They had originally hoped to use this transplantation method with their synthetic *M. genitalium* genome, but they grew impatient with this cell's slow reproduction rate (Venter, *Life* 111–26). They switched to their previous focus, *M. mycoides,* for future experiments (Gibson et al., "Creation"). Once again, they hoped to transplant its genome — synthetically constructed this time — into *M. capricolum*. An unanticipated difference arose; their synthetic genomes seemed to be digested by their host cells, a problem they had not faced with natural genomes. They concluded that these two species must have similar methylation patterns in their genomes (an "epigenetic" marker that does not change the nucleic acid sequence), the absence of which in their synthetic chromosomes was causing the transplant to be rejected. By methylating their synthetic genome, they successfully created a reproducing cell line with a functional synthetic genome, which they referred to as JCVI-syn1.0. It was greeted in the press as "Man-Made Life" or "Man Made Life" (Madrigal; "And Man").

Ultimately, their goal was to use their ability to synthesize and implant genomes in order to create a cell with a "minimal" genome. The concept of minimality is one I will place in question in a moment, but methodologically it meant that they deactivated genes from a cell's genome in order to search for the minimum set of "essential" genes with which it could still sustain itself and reproduce. First, they disrupted the genes in *M. mycoides* using a process called transposon mutagenesis, and deemed genes inessential if cells retained their vital functions after those genes were deactivated (certain genes that were not necessary to reproduction, but impacted the growth rate of the cell, were deemed "quasi-essential"). Now, many genes are dispensable when deactivated one-by-one, but if deactivated in unison with another gene, prove "essential." In order to find a minimal set of genes that would still function after the removal of hundreds of genes from the genome, it was necessary to go through a painstaking process that minimized one segment of the genome at a time. They succeeded in reducing the *M. mycoides* genome to 531,000 base pairs, smaller even than *M. genitalium*'s genome, and implanted it in an *M. capricolum* cell, which remained viable. Venter described this "minimal" genome as nothing less than the essence of universal life, and a step toward life's rational design from "first principles" (Hutchison et al., "Design" 8).

Venter has certainly developed a mythos around his research, and one that can be placed in question. That is not to say that the real of life and science lies elsewhere, in another language. Rather, by placing on stage the artificiality of these constructions — not their synthesis from first principles, but their dependence on an alterity that cannot be grounded in a simple origin — I hope to show that there is no life or science prior to such alterity. We are not recovering a pure life or pure practical power of engineering from the imposition of a false frame, but recognizing that anything we imagine this "science" can say or do, prove or engineer, with respect to the living, has no existence outside of a deconstructible contextuality.

For instance: if this is a technique, where and how can it be repeated or put to work again — what are the limits of its reproducibility or iterability? Any answer to this question depends on how we interpret these experiments, whether we think they have enacted something on this-life-here or all of life, how we particularize or generalize their effects. Venter tends to give them as broad a scope as possible, arguing that he has answered the very question "What is life?" For example: "Even then, however, we recognized that if we were successful in the ability to design the code of life in the computer, translate it into DNA software by chemical synthesis, and put that synthetic code to work to create a new organism, this meant that vitalism was truly dead and, as a corollary, that we would have a clearer picture of what the word 'life' really meant" (Venter, *Life* 78). Vitalism is "truly dead," and "life" is just this death. The *truly* here is obviously symptomatic — Venter knows the history of his discipline well enough to know that he is not the first to put life to death. Nor is he the first to think he would be the last. Already, in 1912, Stéphane Leduc could write that his research had demonstrated the same: "Living things are made of the same chemical elements as minerals; a living being is the arena of the same physical forces as those which affect the inorganic world" (*Mechanism* xiii). And yet, Venter refuses to even acknowledge this predecessor, and after several generations of identical claims he still feels that this needs to be proven again-for-the-first-and-last-time. What no science, theory, or philosophy *as such* could think, what disobeys the order of rational thought and "first principles," would be a "life" that only was or first began as this *coming back*, a *revenant*. What received its death sentence always and everywhere, yet never remained purely dead, never present but never simply absent, requiring a deathblow but never being there to receive it, offering only substitutes and automata. If a pure machine would be something known from first principles, thus exhaustively controlled and programmable, while a pure life would be a self-possessed will, a programmer, neither has ever existed. Everything we piece together here and there, bricolage and bricoleur, is as if it were caught in transit between the two, as if we received the life of the other.

What is the truth of life that Venter believes he has, at last, revealed? A certain tension is observable in his descriptions of this apocalypse — if we are to be rid of the life of the vitalists, to receive new life, reborn with full self-knowledge, redeemed of its original sin, clearly, he will have to offer something *new*. And yet:

> The other major impact of the first genome transplants was that they provided a new, deeper understanding of life. [. . .] DNA was the software of life, and if we changed that software, we changed the species, and thus the hardware of the cell. [. . .] Our experiments did not leave much room to support the views of the vitalists or of those who want to believe that life depends on something more than a complex composite of chemical reactions.
>
> These experiments left no doubt that life is an information system. (Venter, *Life* 109–10)

On the one hand, he promises "a *new*, deeper understanding of life," and on the other, he offers the cybernetic representation of it that has been biological received wisdom for about seventy-five years. Venter is quite conscious of the history and prehistory of this representation; he gives a sort of hagiography of Schrödinger's *What Is Life?* lectures, and many of his descriptions could have been copied from Richard Dawkins or countless others: "Life ultimately consists of DNA-driven biological machines. All living cells run on DNA software, which directs hundreds to thousands of protein robots" (Venter, *Life* 6).[8] Life as "information system" implies that a structured system of differences, like the binary logic gates of a computer or the quaternary alphabet of DNA, could execute a set of programs to process inputs and output the behaviors we observe as life. Despite describing that program or that information as if it were present from the beginning, and as if all of life was derived from it as output, we the living are necessarily in the position of trying to recover programmaticity from a derivative without origin. We bear witness to something several programs, perhaps a nonfinite number, could have created, and are left to search out the true law — "we" too are one of the effects of such programmaticity. Vitalism and mechanism are complicit in this dissemblance of the nonorigin; one treats life as the decision of the spontaneous or autonomous will, one as the inexorability of natural law, but in both cases one must posit a self-identical power of origin of which all of life is the manifestation. If simple and self-present origin is impossible in principle, and yet life and anything else only ever takes place as if it had been there, one can account on the basis of this structure (necessary-and-impossible-origin), for the pattern we have been observing here:

"life" as the recurrence of a putting-to-death, what cannot stay dead because it has never simply lived. It takes place between the "no doubt" and the "not much room" Venter speaks of leaving open for vitalists. How much is not much? Is it none? Very little, almost nothing.

If there is something idiosyncratic in Venter's cyberneticism, something that might merit the claim of novelty, it is perhaps his description of DNA as "the software of life."[9] He's not the first to deploy this analogy, but he may put his own spin on it (Moralee). If DNA-as-software is less common than general invocations of life-as-information or program, it may be for good reason; it quickly leads Venter into strange contortions of figuration. In the passage we just read, this software is granted the power to create the hardware that reads and executes it: "DNA was the software of life, and if we changed that software, we changed the species, and thus the hardware of the cell" (Venter, *Life* 109). Elsewhere, Venter refers to synthetic genome transplantation as "the genetic equivalent of figuring out how to run PC software on a Mac" (*Life* 111). Just a metaphor (or two), of course; nothing to take too seriously — and yet. This is not typically how software, or anything else, functions. It would be an unusual machine indeed that could programmatically construct its own hardware, but it would be even more unusual for it to also be a *virtual machine*, an operating system running on another operating system, (at least) two degrees removed from the hardware. Now, one might be able to link these representations by saying that an old cell's "hardware" was necessary to "boot up this new software of life" (Venter, *Life* 130), at which point this software was empowered to reconstruct that hardware. Throwing away the ladder it had climbed, so to speak. However, any attempt to give rational form to this figuration must acknowledge what everything in Venter's discourse seems designed to suppress: the dependence and conditionality of the "DNA software of life" on what it is not. We will see in a moment that this dependence does not even stop at the membrane, where Venter would likely draw the line of what could reasonably be designated "hardware," but includes the environment or artificial medium. His metaphor is all the more ironic in that it is this contextuality that determines what information the "software" could be said to contain. Not just whether it can be read but how it is read, as what language and so on. Venter would like this image to recapitulate what Schrödinger described as life's "law-code and executive power" but is demonstrating despite himself that these roles are internally divided (Schrödinger 22). It is not that software or legislative-executive power resides somewhere else, or can be represented by a better figure, but that there is ineluctable figurality because one never simply returns to a unified source of power. Its internal differentiation appears as the excrescence of so many inessential figures.

To elicit the effects of contextuality overdetermining genetic agency, I will examine the concept of the "minimal" genome that animates much of Venter's search and research. One could find more than one reason to attempt to reduce the number of genes in a still-viable cell, but again my hypothesis or obsession here is that there is no description or even performance of such an experiment that is not always already *reasoned.* It is motivated by or drawn towards more than one or perhaps a non-self-identical origin or telos that may be supplantable, but the "experiment" itself or the "reality" itself has no being prior to its animation or division by what only after the fact can be parsed as the interpretation of a fact or a given. It cannot be, then, to recover a literal truth that I question Venter's habitual claim — almost everywhere that he discusses "the" minimal genome, it is treated as the minimal genome *of life*. Not of this life here, of this species or this individual (in which case, what reproducibility could we expect of our experiments?) but of *life in general*, as if there were such a thing. He justifies his experiments as making it possible to "see and understand the minimal gene set for life" or to understand "which genes might be essential for minimal life" (Venter, *Life* 85). If I would place in question this self-understanding, it is not to locate the essential or minimal elsewhere, but to say that there is no "essential" or "minimal" except as the effect of some contextuality.

This is acknowledged on several occasions in Venter's research, as it must be, though certain necessary consequences are never drawn or are stubbornly resisted. Venter's team explains their research by arguing that their "minimal" genome represents only "essential" genes that perform "nearly universal functions" (Hutchison et al., "Design" 10). Whatever differences exist between their scientific papers and Venter's writings for a popular audience, this rationale remains unchanged across contexts; it is as if there were an itemizable set of vital functions shared by all living things, that could be mapped onto individual genes (each gene contributing to one of these functions). What becomes of this representation when context must be taken into account?

> A minimal cell is usually defined as a cell in which all genes are essential. This definition is incomplete, because the genetic requirements for survival, and therefore the minimal genome size, depend on the environment in which the cell is grown. The work described here has been conducted in medium that supplies virtually all the small molecules required for life. A minimal genome determined under such permissive conditions should reveal a core set of environment-independent functions that are necessary and sufficient for life. Under less permissive conditions, we expect that additional genes will be required. (Hutchison et al., "Design" 8)

First, the problem: we would like to locate all and only what is necessary to the cell, but what is intrinsic depends on the extrinsic. A certain gene will be dispensable in one environment, but not in another (and a possibility that goes unmentioned here: dispensable for one cell in one environment, but not another in the same). If we still hope to believe in the promise of essential life, it is necessary to imagine that there is a medium or mediator that could act as a sort of plenum, in short a *mesitēs* or *logos,* purely communicating the essence of life. The "permissive conditions" of a "medium that supplies virtually all the small molecules required for life," a kind of Eden or paradise regained, would reveal at last the original essence: "a core set of *environment-independent* functions that are *necessary and sufficient* for life" (emphasis added). Instead, what one finds is a sort of supplementation, one less is one more, and rather than returning to the minimum or simple one constantly re-invents complexity. For instance, Venter's team found that their cell, because of its enriched artificial medium, could do without genes that other cells use to manufacture amino acids, nucleic acids, sugars, and so on. However, because it was dependent on its medium for what it could not create itself, they found that an unusual number of its "essential" genes were involved in transportation across its membrane.[10] Though some have gone as far as to argue that the "minimal genome" concept should be subdivided into "essential genes" and "context genes," precisely what we are witnessing is the undecidability and instability of any such distinction (Simons 132).

The contextuality of "essence" is well captured by a curious detail about *Mycoplasma*, the genus of bacteria Venter's team has used as a starting point to search for a minimal genome. *Mycoplasmas* have a long history of being sought out as representatives of minimal life. Perhaps the first researcher to do so, Harold Joseph Morowitz, hoped that their stature and simplicity represented a proximity to the origins of life (Simons 128). He assumed a connection between the original and the essential, the essence being what was present at the origin. As it increasingly became the consensus view that *Mycoplasmas* were "degenerate" bacteria, that they had lost size and genes during their evolution as parasites, Morowitz adamantly resisted this possibility: "the mycoplasma is primitive" (qtd. in Simons 129).[11] It is now agreed that the "minimality" (such as it is) of *Mycoplasmas* is a latecomer. The second term in an organism's species name often refers to the location where it is found: *M. capricolum* (from Latin, goat-dweller) is a parasite of goats, *M. genitalium* causes a human STI (Venter notes that when his team synthesized its genome, they deactivated a gene that allows it to infect human cells).[12] *M. mycoides* has a history that is of particular interest given the themes we are tracing here; it is the cause of bovine pleuropneumonia, and played a curious role in the earliest debates over the nature of the "filterable viruses." In 1898, it stood on the very cusp of filterability (it was found to pass through the Pasteur-Chamberland filter only under certain conditions) and of visibility by microscopy — but unlike other filterable viruses it had been successfully grown in culture (Hughes 64–67). This microbe of "extreme tenuity," suggesting the possibility of others "beyond the limits of visibility," challenged the distinguishing marks of the "filterable viruses," leading some to expect they could all be bacterial in nature (Nocard and Roux 248).

Venter's team spins this parasitism in a curious way, to try to restore the value of essence that Morowitz identified with the original. Essence, not as origin, but as end:

> [W]e set out to construct a minimal cellular genome in order to experimentally determine a core set of genes for an independently replicating cell. [. . .] *M. mycoides* has several advantages for this purpose. First, the mycoplasmas already have very small genomes. They have evolved from [. . .] bacteria with larger genomes by losing genes that are unnecessary in their niche as mammalian parasites. They are already *far along an evolutionary pathway to a minimal genome*, and consequently they are likely to have fewer functionally redundant genes than other bacteria. (Hutchison et al., "Design" 8; emphasis added)

Mycoplasmas achieve their relative minimality because they are parasites, and have undergone a process sometimes called "regressive evolution," though the implication this terminology gives of return to a prior state is erroneous. They gradually abandon parts as they come to rely more completely on their host. Even if Venter mostly avoids any naïve speculation that a minimal cell would represent something closer to the origin of life,[13] or the "Last Universal Common Ancestor," he retains a teleology that places simplicity closest to the essence. *M. mycoides* is "far along an evolutionary pathway to a minimal genome," as if it were traveling toward a destination, and this pathway has consisted merely of the loss of "redundancy," of the inessential. On the other hand, Eugene Koonin, who has also written on the minimal genome concept from the perspective of his work in comparative genomics, acknowledges that parasitism can be inventive: "*M. genitalium* [. . .] was dubbed 'minimal' at the time of its sequencing. However, there is no proof of its actual minimality and, moreover, a parasite might have extra genes for interaction with its host" ("Comparative Genomics" 128). Parasitism is not necessarily leading us either toward or away from a pure, autonomous, and creative life — it compromises the notion of essence presupposed by these doctrines.

Whether it is located at the origin or telos of life, the notion of essence plays a similarly anchoring role for the biologist. A set of nearly synonymous values gather around it that is reflected in the language and reasoning of Venter's texts: *essence* is the necessary-and-sufficient, the universal or common, the basic (the simple or simplest possible, without adornment), the return to first principles, and of course the minimal. The essence of life would be what is proper to it *qua life*, thus shared by everything that counts as living, nowhere lacking among them, and would be the *cause* of the life of the living, what makes and gives life and thus can serve the scientist of life as a valid explanation of vital phenomena. It is essential to the essence that it belong to something *of itself*, that it secures the innateness or intrinsicness of the proper and the ownmost. Something accidental, extrinsic, improper, or corrupt, hybrid, admixed would be the inessential, it would depend on something other than life itself, or this life here, to be or become present or absent. If we find certain scientists tying themselves in knots to try to shelter this value of essence, to prevent the improper or inessential from intruding upon it, and to preserve their privileged access to it, it is undoubtedly because what we call science demands recourse to this value. It is the essence of science, though there may be no non-science in this respect (nothing that is without referral toward essentiality). For instance, the role that essentiality plays in Venter's project is not in the least something circumstantial, that could be excised from his texts while leaving the core certainties of his work intact. It is not just to aggrandize his discoveries that this chain of minimal-essential-principial-universal-necessary appears, though it certainly plays a role in something like self-promotion; that something fundamental *to life* has been discovered validates the publication of this work in a journal with a title like *Nature* or *Science*, as opposed to, say, some dusty corner of microbiology. One could imagine a less savvy researcher doing all this work (though they would never have gotten the funding in the first place) just to announce that they had discovered "A Possible Genome Reduction of *M. Mycoides* in Artificial Medium," and receiving far less fanfare.

Has something happened? An event? The most one can say is that, if so, it is never purely separable from its inscription, the reproducible or iterable form in which it necessarily appears. One is free to say that it has been "communicated" wrongly and that the "true" event lies elsewhere, but only by re-inscribing it within another iterable series, and the event remains a question for us precisely because it receives its first chance from this originary re-inscription.[14] One can write that something has been discovered that is valid of "life itself," or of a particular genus or species of the living, of a particular genotype, perhaps restricted to a certain environment, but nothing can be written, said, thought, or experienced (even by the most "empirical" gaze) without some circumscription by iterable categorization. If one were to simply give the object of one's research a proper name — and some have called Venter's monster "Synthia" (Kwok, "Genomics") — then there would be no promise of reproducibility to one's results or one's knowledge, and thus no clear indication of their scientific merit (except perhaps to the extent that they could be taken as an example of something generalizable — a thematization left to their reader). Then, even if it is fairly easy to puncture inflated claims about discovering the essential-minimal-universal of all *life*, one can only do so by invoking a displaced essentiality.

Far from justifying the reign of essentiality, its necessity is what undermines its authority. The essence of essence is the inessential, that it never appears and has no being except within some context, some environment, some extrinsic accident or accretion. One name for this inessentiality is parasitism or virality, and one finds its effects at work within every search for the essence of life. Venter's "minimal" genome project claimed that a certain parasite was on a pathway toward a simplicity representative of universal life, but we can decipher more restrictive and necessarily arbitrary limits around it. In short, this "pathway" to a minimal genome requires an asterisk, about which other authors show more clarity: "The smallest prokaryotic genomes sequenced to date belong to species not considered autonomously alive, which, while missing essential genes, became entirely dependent on much more complex hosts:

insects. '*Candidatus* Carsonella ruddii' has an impressive 160-kb genome, and '*Candidatus* Hodgkinia cicadicola' has an even smaller one, with 144 kb, which leaves scientists at the edge of considering them organelles, as in the case of mitochondria and chloroplasts" (Xavier et al. 489). The parasite or symbiont shares a conceptual edge, a border, with the organelle, such as the mitochondrion, which represents an ancient, endosymbiotic association. The mitochondria in human cells retain just 16kb of DNA (kilobases, or thousands of base pairs — Venter's minimal genome was 531kb), DNA that is more closely related to that of certain contemporary species of bacteria than to that of their hosts (Anderson et al.). If Venter and his associates do not consider this extreme simplicity to be on the pathway to the minimal genome, it is because of an only infrequently stated assumption, crucial to their concept or preconception of life. The "minimal genome" must be able to support an "independently replicating cell that has been grown in pure culture" (Gibson et al., "Complete Chemical Synthesis" 1215). As much as a scientist might hope that the *reproducibility* of such conditions of growth promised the essentiality of the knowledge so produced — its intrinsicness to something *independent* and *pure* — this is only another context; in fact, one that is quite rare and artificial. If this context is so taken-for-granted that it is rarely even mentioned in Venter's writings, that is not because it is essential to life, but because it is common practice among one very peculiar species: the biologist. Microbes are frequently studied in such "pure" cultures, and often granted representative status if they thrive under these conditions, even though it is thought that only one percent of the bacteria on Earth admit of being cultured under these laboratory conditions.[15] Laboratory life, even when its media offer the most "permissive conditions," is a rare breed, contrived to live in a context that we can recognize as thoroughly artificial — even if artificiality may be the nature of nature.

The limits or de-limitation of contextuality work within not only the living but the science of life as well. It is why we are continually left to tinker with models, which are held apart by an unstable spacing from a being, reality, or truth that never ultimately arrives. Certain "independently replicating cells grown in pure culture" are what are often called *model organisms*. Because a species such as *E. coli* is particularly amenable to being studied in the laboratory, a doctrine of hyperspecific knowledge has developed around it, not necessarily because we are especially interested in this particular species, but because it allows for a common practice and language among researchers, and is taken as representative of broader classes of the living (not exhaustively, but one can extrapolate something about its genus, something about its kingdom or domain, something about life in general). In the early twentieth century, as laboratory experimentation became more central to biological study (Kohler 286), model organisms such as *E. coli*, the fruit fly, or house mouse were chosen in part because they did not exhibit developmentally plastic responses to their environment but seemed as much as possible to carry their nature or essence within themselves (Gilbert 207; Gilbert and Epel, *Ecological* 491–92). This *model* of life, as exhaustively determined by internal "genetic" factors, is more in question than ever today, and laboratory practice changes with it. What is sometimes called a holobiont or metaorganism, which includes an organism as traditionally defined together with its symbionts, is now taken as a focus of study — for example *E. scolopes*, a squid that forms a lifelong association with bioluminescent bacteria (Nyholm and McFall-Ngai). Part of the environment, of the context, is in this way brought inside, domesticated, still seen as a symbiont but no longer as thereby extrinsic. This domestication does not eliminate essence or its artificiality but displaces and reinscribes it — the subject of reproducible knowledge becomes the holobiont (under certain conditions, and so on). By suppressing the necessary artificiality of their contextuality, Venter's team performs an all-too-common operation, mistaking the arbitrary limits of their experimentation for the limits of life itself (Simons 133).[16]

Nor is this the only modeling-effect that shapes minimal genome research. Every aspect of Venter's writing reveals its dependence on a concept of the "gene" and genetic function that can be placed in question in multiple respects. His assumption is that each gene, regardless of the others and its environment, contributes primarily to a single vital function, and that the genome in isolation exhausts the causality and identity of the living. For instance, in addition to identifying a set of genes that would count as a "minimal" genome, Venter's team assigned each gene to a single "functional" category (Hutchison et al., "Design" 5). Some of these functional correlations may be quite robust, but it is always possible that careful scrutiny might reveal effects of their contextuality. It may be that a given gene and function are correlated — that a particular function ceases when that gene is disrupted — in a given context, while in another context one might find that function unaffected, or a different function affected. This contextuality certainly opens the "essential" to the "environmental," but also to its *internal* environment. The piecemeal (or beanbag) fashion in which an individual gene can be assigned a particular function, by seeing what happens or doesn't happen if it is silenced, ignores the possibility that a more holistic set of functions may be following from collective or networked interactions of the entities we call *genes* (possibly together with other material structures). It would follow, not simply that the "function" or functions of a given gene depend on other parts of the genome or cell, but that an entirely different model of "minimalization" needs to be considered. Venter does investigate the possibility of "synthetic lethals," pairs of genes that can be individually disrupted without affecting viability, but that kill the cell when silenced together (Hutchison et al., "Design" 4). However, he entertains this possibility only within the limited frame that assumes these two genes must each perform the same function, or slightly different but substitutable functions (he gives the example of two genes, one for transport of glucose and one for fructose, which were synthetic lethals as long as the artificial medium contained both sugars; Venter, *Life* 60). It is also possible — at least, it cannot be ruled out a priori — that much broader networks of genes (and possibly other genomic or cellular elements) may enable cellular behaviors in a relatively holistic fashion. Beyond challenging the Venter team's descriptions of genetic function, this could also entail that much larger groups of genes are substitutable. Even for this specific species of cell, Synthia may be only one of many possible "minimal" or near-minimal genomes (Venter's team considered only the possibility of pairwise interactions — a full combinatoric analysis could reveal other possibilities). In general, Eugene Koonin is more willing to acknowledge the difficulties facing the concept of minimality (though he tends to treat them as pesky though circumventable accidents) — in his words, "This creates enormous combinatorial possibilities for constructing (theoretically, but eventually perhaps, experimentally) numerous versions of minimal gene-sets, even for the same set of conditions" ("Comparative Genomics" 129).[17]

The genetic determinism presupposed by Venter's methods and texts is another way of suppressing the contextuality that makes "essentiality" possible and impossible. No essentiality except where "context" can be bracketed and abstracted, and thus there is only essence as an effect of the iterability of contexts. What arrives as if from outside has always already haunted the "inside" as an inner difference. Nothing about the Venter lab's experiments tells us what would happen if this genome was implanted in a different "species" of cell.[18] In fact, at the same time that a certain model of life (heavily invested in genetic essentialism) motivates these researchers to search for a single set of genes representative of all life, an explosion of genomic data is revealing how little is universally genetically shared. Advances in sequencing technology have made these data available only in the last two decades (Venter's work has played a part in this) and have quickly challenged existing conceptual models. There are only a handful of genes that researchers now expect to find shared across species, and any question of what genes are universally present depends on a prior decision as to what cells we will include in our count — ever smaller symbionts are being sequenced and make the question of essence descend from an arbitrary decision (Lagesen et al.; Land et al.). What is universal among life depends on how we define life, not vice versa. Surprising amounts of genomic diversity are even found *within* species — the most sequenced genome by far is that of *E. coli*, and only two thirds of its genes are found to be broadly shared — the term *pangenome* has been introduced to meet the newfound need to describe intraspecific genomic differences. *E. coli* have on average about 5,000 genes, yet after sequencing 2,000 *E. coli* genomes, 89,000 gene families have been found (these are not genes differing by point mutations, but unrelated genes; Land et al. 141). It remains to be seen how easily one can treat this diversity as inessential — it may not be that only the broadly shared genes are necessary for life, or even that genetic differences can meaningfully be collapsed within categories of universal vital "functions."

Venter's team was quite careful to use a species they knew well and had used as a host for past transplants of similar genomes. If this genome could not function in a different cell, or a different cell could not function with this genome, it would certainly not be possible to declare this genome minimal "for life" in general, but it would also be necessary to ask what besides the genome is "essential" for vitality or viability. Not just what does a given cell contribute, but what does it require, what does it demand? Venter's assumption that the executive or programmatic power of life resides entirely in the genome excludes other possibilities that are increasingly well-recognized today. He refers to his creation consistently as a "synthetic cell" despite his synthesis efforts only grappling with the genome.[19] What effects a gene can be observed to have often depends, in certain contexts, on epigenetic markers and plastic relationships with the biotic and abiotic environment. Even in order to get their synthetic genome accepted by its host cell, it was necessary for Venter's team to methylate it — methyl groups that bind to DNA and change how it is expressed are one form "epigenetic" inheritance can take. Nothing in Venter's method can tell us what role structures outside the genome, including other DNA structures such as plasmids and viral episomes, play in the "functions" of life. These factors that are external or contextual to the essence as Venter defines it (the genome), do not just complicate one version of minimality, but disrupt essentiality itself — if essence cannot be attributed to anything present in or possible for a cell or entity from its "birth," then "life" has no originary power of essence.

There are also differences of modeling that, so to speak, intersect obliquely with all those introduced thus far. The sort of intuition or sensitivity that we are accustomed to attributing to ourselves may well belong to the sort of life we are investigating here. Rather than uniting causality in a vitalistic spirit, this further complicates or subdivides its possibilities. As an analogy: it is possible that a person with a particular brain lesion loses the ability to feed themselves, but we would not therefore conclude that this was the part of the brain "for" metabolism. The genome may be a sensitive organ in an analogous fashion (McClintock). Whether it is pictured as mechanistic or aesthetic (feeling), what we call nature or life involves an iterability that makes mechanical law and recognition ultimately undecidable — it is only by something like an act of reading that the cell forms alliances and disalliances, and welcomes certain substances while rejecting or breaking down others, in that it can only be identified as will or law if it repeats, and thus it is necessarily open to an immemorial past and a future that has never finished arriving. This *natural lection* is the true subject of study for the natural scientist, who thus is only ever engaged in the reading of an act of reading — like all the living, like all of *life-science*. It follows that what life will be *for us* is not the same as what it is *for this life here*, a dissension that haunts every exploration of essence. Among other things, including the philosophico-theological inheritance we have recognized behind the scientific impulse, it is our all-too-human ends that circumscribe certain expectations or limitations expected of life. For example, the "quasi-essential" genes included in Synthia's genome that allowed for faster reproduction were included largely for the sake of the experimenters, and the desirability of this speed for other biotechnological applications. There is only life *for life*, and this has always made me suspicious whenever it is said that one fine day Darwin banished teleology from biology. Whether such a synthetic cell would recognize itself, would accept this as its life, is a necessary question even if it is not straightforwardly answerable. What differences could we have introduced that we do not even know how to look for?[20]

Parent, Patent: Life in Venter/Venter Capital

> Nor could we be sure that its deconstructive structure cannot be found in other texts that we would not dream of considering literary. I am convinced that the same structure, however paradoxical it may seem, also turns up in scientific and especially in judicial utterances, and indeed can be found in the most foundational or institutive of these utterances, thus in the most inventive ones.
>
> — Jacques Derrida, "Psyche: Invention of the Other"

One can already see, then, that no matter how "theoretical" these enframings may appear, they cannot simply be discarded to arrive at a purely separable "practical" core. In fact, Venter and his collaborators' justifications of the minimal genome project are continuous with an often invoked pragmatic motive. The simplification of the genome is sometimes described as progress toward the maximum efficiency and predictability of the bioengineering ideal, and thus as a contribution to the synthetic biology project of turning the cell into a factory:

> With increasing knowledge of the functions of essential genes that are presently unknown, and with increasing experience in reorganizing the genome, we expect that our design capabilities will strengthen. The ability to design cells in which the function of every gene is known should facilitate complete computational modeling of the cell. This would make it possible to calculate the consequences of adding pathways for the production of useful products, such as drugs or industrial chemicals, and would lead to greater efficiency in development. (Hutchison et al., "Design" 10)

If anything, this statement from Venter's team is more cautious than many similar celebrations of the *standardization* of the minimal cell (Simons). This particular claim of progress toward the calculability of the cell comes from the scientific paper published in *Science* in 2016 by the JCVI team that had just succeeded in synthesizing and transplanting a "minimal" genome. When Venter speaks in more public-facing contexts, he can be even more deliberate about emphasizing the commercial possibilities of this project, and making his pitch to investors:

> DuPont is a world leader in using modified bacteria on an industrial scale to transform renewable natural resources into products that would normally have to come from oil. They are building a \$100m plant in Tennessee that will turn corn sugar into propanediol, a key polymer for making plastics, but it has taken them over five years to modify a bacterium so that it will turn the sugars into the polymer efficiently. Most of the work involved shutting down existing pathways in the bacterium so that the sugar would end up as the polymer and not be used for something else. The result showed that huge gains in efficiency are possible. My approach starts from a different place. I think that if we can build a cell from scratch with only the very minimum of processes it needs to survive, we won't need to go through this long process of modifying an existing bacterium to shut down all the pathways you don't want. Instead, you can add to this "minimal cell" the pathway you need in order to make a specific product. I have been trying to understand the minimum a cell needs to survive for the past ten years, as part of a basic research project; the fact that it has commercial and social applications is wonderful. (Anderson)

My analysis above may have seemed to relate most closely to what Venter calls his "basic research project." Questions about the definition of life, the species concept, mechanism and vitalism, genetic determinism, and contextuality might seem to focus on theoretico-philosophical concerns that, if they relate to scientific research at all, do so only at the level of ideological justification masking its economic imperatives. If Venter has been able to mobilize vast amounts of funding to complete this "basic research project," it is not because some philanthropic and philosophical soul expects it will answer the above questions, but because investors focused on the commercial arm of his scientific operation want a stake in the technology and intellectual property emerging from it, or because the government has its own interests in these potential applications. Nonetheless, if one examines his research and its justifications from this commercial perspective, every apparently theoretical concern recurs at the heart of its economic and legal motivations.

When recognizing the cross-contamination of relatively theoretical and practical contextualizations, one must avoid thinking of those questions in a philosophical form as first principles from which practical effects are derived. Rather, it is because nothing with the force or integrity of a first principle or simple origin can be found that each of these contexts encompasses the other.

There is no purely reflective space in which questions about nature or representative natural entities such as "the cell" can be broached except in the context of the massive apparatus of the natural sciences, a machine overdetermined by every political, economic, technological, and cultural concern circumscribing its functions. Even the space for quiet contemplation sometimes granted to certain individuals is a pocket carved out of this device's circulations, a reserve from which it often finds ways to draw profits. At the same time, and for the "same" reasons, there has and will never be *purely* practical, economic, or political motives, if this is to mean that they can be distinguished from the theoretico-philosophical or questions of "basic" as opposed to applied science and made to purely determine the latter. Every undecidability that sutures apparently theoretical reflection divides practical power and its political or legal protections from themselves. Any attempt to demystify "ideology" or to focus on "material" conditions will be exposed to an unconscious in which specters of an ontotheological, Aristotelian or Abrahamic, inheritance haunt any effort at circumscribing the pragmatic. In other words, precisely because there is no grounding theoretical principle from which the proper explanation of Venter's results follows by logical derivation, and yet there is no way to undertake a technoscientific description at all without the pretense of such rationalism, there will always be something in excess of apparently immediate, practical, or technical concerns that may seem to dictate the workings of the econo-techno-scientific machine as if by remote control. Venter's descriptions are in a sense bad science and even bad engineering (they could, taken at face value, lead to false expectations or predictions about the generalizability of his results, or how such creations could be put to work in other contexts, among other things), but that does not mean that there is any other way to describe them that is free of such criticizability. So, we are left to ask why this description was not only preferred by Venter but proved sufficient to set capital flowing in his direction. I am not denying that some of this efficacy seems obvious (promise control, promise profit, promise it sooner rather than later), but not all — and that is what sets me wondering.

Venter's career offers a kind of object lesson for the circulations of these forces and their undecidability. He first rose to notoriety (indeed, he is often referred to as the "bad boy" or "rude boy" of biology) during the Human Genome Project, when, as a government researcher with the National Institutes of Health, he was associated with the filing of patents on a thousand genes (even prior to their sequencing, merely on the short expressed sequence tags that were used to identify the location of genes). At the time, he argued that securing intellectual property rights would encourage private companies to develop medical applications from this information. He was denounced by many others involved in the publicly funded genome project, which may have been just what he hoped to accomplish; this first skirmish launched his reputation as the "disruptor" of a torpid governmental-academic bureaucracy, ultimately leading to his work with the private sector. First, Venter was offered seventy million dollars in funding for a nonprofit of which he was president (The Institute for Genomic Research or TIGR), in exchange for which a for-profit company, Human Genome Sciences, would receive the intellectual property rights to the genes discovered by TIGR. He was then sought out to be the president of a new for-profit venture, Celera Genomics, which announced in 1998 that it would race the public effort to sequence the genome. There were immediate concerns that a private corporation promising to accomplish this task at no cost to taxpayers would lead Congress to reconsider its funding of the public HGP. Celera's business model, based on the privatization of genomic information, also led to concern among many scientists that it would jeopardize the sharing of information on which science depends. The ensuing controversies only further elevated Venter's profile.

Celera's business model demonstrates both the role of intellectual property in the economy of biology, and the speculative circulation in which IP and biotechnology are mere nodes. It promised to become profitable both by selling subscriptions to a database of the genomic information it was sequencing, and by seeking patents on some of the genes it discovered in the process. The plan befuddled many onlookers and even most of Celera's employees and executives (Shreeve). The value of its database was in constant tension with Venter's desire to achieve scientific recognition, which required making its data publicly available in some form. Moreover, by seeking provisional patents before revealing its sequence data to database subscribers, it undercut the database's value for its commercial customers (drug companies looking for the same IP).[21] The real reason that it attracted the attention of investors was its ability to promise big — the unveiling of the genome was viewed as an informational gold rush — and Celera was promising to get there first. While Celera's work may have ultimately advanced the limits of what could be done with certain sequencing technologies (the "whole genome shotgun"), it created a kind of race to the bottom that led both the private and public ventures to declare mission accomplished after only completing partial "drafts" of the genome. Like many "disruptors" since, while Celera built a mystique around being more efficient than the public sector, its plan to lock up public goods as IP ended up relying on the publicly funded and publicly available data from the Human Genome Project (which it made use of to finish its draft). Celera's stock skyrocketed after a write-up on investment website motelyfool.com, which was perhaps not out of character for speculative capitalism when it recommended investing despite acknowledging that "[Celera] has no profits, no real revenue, and it has no clear business model, *just a bunch of promises*" (qtd. in Shreeve 307; emphasis added). Their share price soared to $247, fueled solely by these promises, before plunging to $15 just two years later (part of a broader biotech sell-off that presaged the dot-com bubble burst), from which we can perhaps draw a lesson (Rabinow and Dan-Cohen 11). Even if speculative capitalism takes a particular form when its investment vehicle is information, and biological information, this speculation may be the fundamental form of this and all economies (beginning with that circulation and exchange established by our father who sees in secret). In other words, Celera's business model need not be seen as a failure; regardless of whether the promised profits ever materialized, it mobilized immense investments that made Venter, among others, quite wealthy. In 2002, he was able to invest $100 million of his stock holdings in his not-for-profit, which was continuing to work on the minimal genome project (Shreeve 373). Increasingly, manufacturing such hype bubbles is the deliberate aim of investment capital, as self-driving cars, crypto and non-fungible tokens (NFTs), and now "AI" have demonstrated — both manufactured goods and information are mere relays in a system in which the product is the promise.

After leaving Celera in 2002, Venter maintained a business model that allowed him to pursue "basic" research and its commercialization in parallel (Shapin). As he described it, "The scientific founders of biotech companies are mostly just looking for new ways to get their science funded, and when the company comes under pressure to make products, the founder gets the chop. I was trying to avoid that fate by coming up with a model that would allow me to do basic science and reap commercial benefits" (Anderson). TIGR was ultimately consolidated into the J. Craig Venter Institute (JCVI), under whose auspices the later stages of the "minimal" genome research were conducted. JCVI is the nonprofit arm of Venter's business empire, which also includes the for-profit Viridos (originally Synthetic Genomics Inc. — the name change likely marks a shift from capitalizing on the market share of genomics and synthetic biology to that of green tech). Viridos funds the vast majority of JCVI's research, in exchange for which it receives the intellectual property derived from that "basic" science (Venter, "Conversation" 135). The for-profit company can then use the innovative techniques and the publicity of JCVI's headline-grabbing accomplishments to attract customers (Synthetic Genomics Inc., "SGI Applauds"). Viridos develops synthetic biology-based technologies, such as biofuels or vaccines, in partnership with corporations such as Monsanto, ExxonMobil, and Novartis (Howell; Synthetic Genomics Inc., "SGI and JCVI"; Monsanto).[22]

Thus, the "basic" research conducted through Venter's nonprofit institute is, in several respects, interwoven with the commercial aims of his for-profit venture. The resulting media attention is one opportunity to renew Venter's pitch to investors (by talking up the economic potential of these discoveries), and even the gee-whiz effect that emanates from his philosophical musings about, for instance, the definition of life can be just as conducive to the aura that attracts his business partners. Even the purest or most basic science is reappropriable in the circuit of capitalization. This is already true before we take into account the most obvious source of market value and power derived from this research: intellectual property.

Every aspect of the "basic" research discussed above has been the subject of one or several patent claims by researchers affiliated with JCVI. Moreover, every question that disrupted the smooth functioning of the theoretical apparatus necessary to interpret this research rears its head again in the domain of intellectual property, with inevitable practical-economic consequences. Each of the three stages Venter delineated in the outline of his minimal genome project — minimization, synthesis, and transplantation — has been the source of a patent claim. While some of these claims were rejected or abandoned, JCVI has sought intellectual property covering "Minimal bacterial genome," "Synthetic genomes," and "Installation of genomes or partial genomes into cells or cell-like systems."[23] This same research has also led to patent claims on "DNA assembly, methods for cloning and manipulating genomes [. . .] [and] methods of DNA synthesis" (McLennan 258). Now, the question a patent poses or responds to is the same as that question that most haunted Venter's attempts to achieve a theoretico-scientific grasp on what was happening in his lab: To what extent, under what terms or in what contexts, can one control the reproducibility or iterability of a subject matter?

If we thought that speculation or dissemination would cease when we entered the courtroom, we would be mistaken. At first glance, the juridical discourse surrounding JCVI's patents represents nothing more than typical legal wrangling. Of course, there is incentive for Venter's corporations to seek the broadest possible patents, and so it is no surprise that their initial claims are frequently rejected on standard grounds. For instance, the initial filing for a patent on "Installation of genomes or partial genomes into cells or cell-like systems" would have given JCVI and SGI intellectual property over the transplantation of any genome, "man-made" or "naturally occurring," into any type of cell or even a "cell-like system," which could cover all sorts of "protocells" or chassis that synthetic biologists are still working to create. Now, a patent can be sought only for an actual invention, not an imagined or expected one, and must "point out and distinctly state," or specify exactly the parameters of the invention it covers. While these inventors claimed, initially, to have created a method for transplanting any genome into any cell, they were required to narrow that claim down to correspond more closely with the experiment they had actually completed: the transplantation of an *M. mycoides* genome into an *M. capricolum* recipient (McLennan 291). There are at least two reasons why this revision is of particular relevance to the questions I have already raised in the vicinity of Venter's research and his interpretations of its results. First, we can recognize this obvious symmetry: exactly the same question recurs in the theoretical or scientific and the legal-economic domain, the question of categorization or generalizability. Had Venter accomplished something or proven something that was, as he often wrote, valid *for life*? Thus, ought he to be granted intellectual property covering a domain coterminous with the living, covering transplantations between any cells whatsoever? And even beyond this, covering any sort of synthetic life? Or had he merely demonstrated something about this cell here, and perhaps only within the local conditions of his laboratory? This invention's legal status, just as much as its scientific status, depends on this taxonomizing generality, and yet — second point — neither the broadest nor the narrowest descriptions of the domain of reproducibility covered by this intellectual property right have any true coherence. It is fairly obvious to point out that Venter's lab has not yet demonstrated the efficacy or utility of its procedure for *all* possible genomes or recipient cells, and certainly not for "cell-like" recipients that still do not exist, yet even the claim for a procedure or methodology, a repeatable or iterable technique, valid for all genomes and cells of these *species* cannot be guaranteed. Not only because nonfinite attempts would be necessary to demonstrate such validity empirically (for all such cells that are and all that will be) but also because, in principle, no one has ever known what a species is, or where its true limits are to be found. Many biologists themselves share an agnosticism about this "concept" (the concept of species, a species of concept); it is even relatively uncontroversial to argue that species names serve a necessary purpose for communication among specialists without being rigorously definable or corresponding to divisions in nature or "natural kinds." I would say, echoing Augustine, that we know what *M. capricolum* is so long as no one asks us. And that may be good enough for what we typically call "science," and for the courts or the patent office, but it leaves every claim to scientific knowledge and practical efficacy or intellectual property insecure. There is, then, something speculative and ill-defined about every claim over the proper domain of reproducibility of an invention, whose delimitations are just as arbitrary in the narrow as in the general case, and which cannot even be said to come to a rest with the arbitrary decision of the judge deciding these questions. Even if a claim relative to these specific species was granted by this court, the amorphousness of the species concept — and all concepts — necessitates that it will only be the next decision of the next judge, the next doorkeeper before the law, that will reveal, for the time being, what is or is not included in that classification or that right.

The attempted patent on "Minimal bacterial genome" raised another question of generalizability, one that echoes the questions synthetic virology confronted us with in the introduction of this book. The 2002 experiment that created "wild-type" poliovirus from a DNA sequence stored in a computer raised an impossible question: what is a virus, and more specifically, is it matter or information? Again, it was not that we tried to answer this question — rather we found it impossible in principle to rigorously decide — but that this very impossibility set in motion the destabilizing effects of a lack of control that overdetermined both scientific discourse and the relatively pragmatic domains it touched upon (defense, epidemiology, politics, etc.). Well then, if one patents a DNA sequence, do they patent matter or information? (Several years after the initial filing of the "Minimal bacterial genome" patent in 2006, the Supreme Court ruled that genes themselves could not be patented, because they were products of nature, but certain artificial DNA sequences could — I'll return to the questions raised by this decision.) JCVI's patent specified its invention not only as the "polynucleotides" contained, for instance, in a "free-living organism," but also as those "same" genes "recorded on computer readable media"

(Glass et al., "Minimal bacterial genome" 7). Once again, we are faced with an overweening desire to control the iterability of what is fundamentally unstable because its "nature" is just this iterability. The effort to control informational reproducibility reaches almost paradoxical heights, as legal scholar Ilaria de Lisa explains: "The mere downloading and printing of the patent documents presented to the Patent Office would constitute an infringement" (de Lisa 173); moreover, according to another legal scholar, Rebecca Eisenberg, such a patent might even result in the "Catch-22" that the Patent and Trademark Office would violate the very patent it granted by publishing it (Eisenberg et al.). In effect, such a patent would prevent even the perception of the patented information, as Eisenberg explained when writing about an earlier patent also related to Venter's work:

> The claim to this sequence in [. . .] computer-readable medium [. . .] precludes others from perceiving and analysing this sequence information itself. This is distinguishable, I think, fundamentally different, from old-fashioned DNA sequence patents that claimed only molecules. You can read those old-fashioned patents, learn what the DNA sequence is, type or scan it into your computer, search for similarities to other sequences in databases, all without infringing. By contrast, this [. . .] capture[s] much, if not all, of the informational value of the discovery, not just its tangible value. (qtd. in de Lisa 173n491)

Different examiners gave different, perhaps conflicting, justifications for rejecting such patents. At least one argued that the "Minimal bacterial genome" patent pertained to "the content of the information," to an abstract idea rather than a product or process (qtd. in de Lisa 173). After the decision banning the patents of naturally occurring genes, examiners seemed to accept the equation of information and gene, in order to reject the patent on the grounds that both information and matter fell within the products of nature exception (one can patent only products of human invention, not products or laws of nature) (de Lisa 174). Regardless of whether we think this decision is for the best, for the sake of science, open inquiry, its economic consequences, some ideal of jurisprudence, because it prevented a "biopolitical" or "biocolonial" occupation of our body proper, or for any other reason, we can already see the artificiality of the categories supposedly governing the decision, including the category of "nature." Both the conceptual chains {information — artificial/abstract — separate from gene/nature} and {information — equivalent to gene — natural} were used as arguments against this patent.[24]

The Supreme Court case that decided "naturally occurring" genes were not patentable, *Association for Molecular Pathology v. Myriad Genetics, Inc.*, made a curious compromise. This case was brought by the ACLU against a company that owned patents on genes associated with breast cancer risk, intellectual property it used to prevent others from testing patients for mutations to those genes. The Supreme Court ruled that, as products of nature, genes could not be patented — but that artificial gene sequences formed in the lab, even ones closely related to "natural" genes, were patent-eligible. For example, an RNA molecule formed by transcription from a gene, with sections known as *introns* excised, can be turned back into complementary DNA (cDNA), which is patent eligible. Myriad was not entirely wrong when, in its defense, it argued that the genes it was patenting were themselves products of intervention and purification in the lab, never "naturally" occurring in such a form. The concept of nature itself is an artifice, which sets all these borders spinning. My point is not that this decision should have turned out otherwise, but simply that everything from the theoretico-scientific to the economic, political, legal, and practical decisions in this domain follows not from the availability of "nature" or its contrast classes but from the impossibility in principle of fixing these conceptual limits, from their deconstructibility.

A judge brings down their gavel and decides on biology, on what will be written in future scientific articles. A scientist publishes in *Nature* the oral arguments for their next court hearing. What we call science or nature cannot take place without this legalistic supplement. For instance, much of the rhetoric we have already seen Venter deploy, even when it most seemed to represent philosophical or idle speculation, has an immediate pertinence to these legal contexts. Venter's grandiose and often misguided descriptions of DNA as the "software of life" just so happen to align with a recently developed framework for intellectual property allowing for the patenting of computer software. Initially, software had fallen through the cracks of existing copyright and patent law, as these legal scholars explain:

> Copyright covers original works of expression, explicitly excluding works that are functional. Patent law requires functionality; however, it had traditionally been understood to exclude formulas and algorithms. Thus, software — a machine made of words, a set of algorithmic instructions devoted to a particular function — seemed to fit neither the copyright nor the patent box. It was too functional for copyright, too close to a collection of algorithms and ideas for patent. [. . .] As a result of statements by the US Congress and actions by the courts, software ended up being covered by both copyright and patent in the US. (Rai and Boyle)

This relatively recent framework allowing for the patenting of software is part of the context of contemporary synthetic biology. Thus, when Venter and other researchers associated with JCVI describe DNA as software (in their scientific articles, trade publications, interviews, etc.), or when they deploy programming conventions such as giving their synthetic cells version numbers (JCVI-syn1.0 and so on), they may well be jockeying for another route to the patentability of their creations (Calvert 172). Legal reasoning often proceeds by just such analogies.

It is frequently difficult or impossible to say which of the several audiences for his work Venter may be speaking toward at a given moment. His claim to have created "synthetic life" certainly garnered headlines and the interest of the hoi polloi. It led to consternation from certain scientists, who argued that because the synthetic genome was derived from a naturally occurring one, Venter was still operating in the domain of mere imitation of a natural product, not the creation of life. (I have argued already that this distinction between creation and imitation, or production and reproduction, which grounds the distinction between nature and technology or art, is deconstructible.) In response, Venter often quips that his synthetic cell is the first organism whose *parent is a computer.* Whatever image Venter hoped to project to the public, and whatever he hoped to prove to his fellow biologists (and perhaps the Nobel committee), we can see that this self-defense also preserves one of the arguments for patentability of his product. Sometimes this emerges explicitly in his self-justifications: "this is clearly the first life form totally developed out of a computer and by humans, so it is much closer to a human invention" (qtd. in McLennan 251). A human invention, not a product of nature — thus eligible for intellectual property protections.

It is too soon to say which if any of these inventions or patents will have been most successful or most lucrative. As the brief (pre-)history of synthetic biology introduced above sought to demonstrate, only the future will tell what has been an invention, who invented the present, and it is not even a question of givens (such that future iterations would wear on their sleeves or display to the pure light of reason what invention or inventor made them possible), but rather it is always a revisable or deconstructible narrative of inheritance or parentage that creates such lineages. Just as, in the case of these patents, there will only ever be an endless forensic antagonism over what has been invented by whom, what possibilities descend from what invention or inventor. It goes as so with all iterability, that the past descends from the future and the concept or principle from the instance. A future that has never finished arriving, and an instance without self-identity.

Without, then, being able to make any once-and-for-all claims, with full awareness that everything may look different tomorrow or the next day and that every word I write and could write is thoroughly exposed to this future, with no reserve or shelter of essence, I will nonetheless hazard one observation. At the present, it would appear that the most successful invention or intellectual property to emerge from Venter's minimal genome project is none of these patents most closely related to what he has framed as its most significant conceptual and technological breakthroughs. It is not minimization, not transplantation, nor is it genome synthesis *per se*. The most widely utilized and likely one of the most profitable techniques to emerge from this work is one its inventors refer to as "Gibson Assembly™" (McLennan 282–83; Cf. Gibson et al., "Methods"). One of JCVI's researchers, Daniel Gibson, was able to streamline the initially laborious process they used to synthesize the half-a-million base pair *M. genitalium* genome into a far more rapid procedure the bulk of which could be completed in a single step ("one-pot"). His patent on "Methods for in vitro joining and combinatorial assembly of nucleic acid molecules" has since been granted, and Viridos has a licensing agreement with New England BioLabs to sell a kit for performing the "Gibson Assembly™" method (Synthetic Genomics Inc., "SGI Announces"). This kit is available for "internal research purposes for the sole benefit of the purchaser only," while commercial uses of the technology require negotiation directly with Viridos (Kahl and Endy 11–12). Viridos even formed a subsidiary, first called SGI-DNA Inc., then Codex DNA, and now Telesis Bio, whose CTO is Gibson and that primarily markets products related to the Gibson Assembly™ process. Its website advertises the company as "Creators of [. . .] the industry-standard Gibson Assembly® method" (Telesis Bio Inc.). In a survey of 137 self-identified synthetic biologists, about half reported that they currently use this method and an additional 22% had used it in the past (Kahl and Endy 5–6). Without being able to assess Viridos's financials in detail, this high percentage is one indicator that this may be the most successful (in several senses) invention stemming from the "minimal genome" research, at least for the time being.

This result reveals something about the nature of what we still sometimes call scientific "progress." Taking stock of this present moment (while acknowledging, again, that it all might seem different tomorrow), it may be that none of the steps Venter fixated on as landmarks of technoscientific advance will be the most influential or transformative legacy of this project. We saw above that when he tried to frame genome synthesis as a conceptual advance, he arrived back at cybernetic clichés. Yet, this innovation in DNA synthesis that seems like a merely technical procedure, a refinement of efficiency for an existing process, could well give rise to fundamental reimaginings or reinventions of biological possibility. Venter argued that minimal genome synthesis was a difference in kind, a qualitative sea change, introducing a novel interpretation of life and its mechanisms. In contrast, Gibson Assembly™ seems at first glance to be mere quantitative change, a process that was already understood as possible simply made bigger, faster, and cheaper. Yet, it is precisely by such redundant or stutter-step differences, which resist conceptual circumscription, that the terrain shifts under our feet without our necessarily being able to bring just how and why to positive knowledge. This lifeblood of science and everything else under capitalism, the efficiency of the bigger-faster-cheaper, often allows for a radical displacement of possibilities. Even something that could be foreseen as possible changes its nature as it falls within the budget of the average lab, the time frame of the average dissertation or grant. Applications of this technology — for example, DNA digital data storage (Hoose et al.) — shift science fiction into active research programs without any grand theoretico-conceptual transformation such as those cathexes of "man-made life" that Venter shared with a broader public and some in his community of scientists.[25] It may create effects of theoretical or conceptual difference, but only because it first exploits the non-self-identity within any conceptual formation. Before we know it, the language surrounding "synthetic biology" may grow stale and the justifications animating many of its marquee research projects may no longer make sense within a new "paradigm," or perhaps will be subtly misremembered in order to assimilate them without reflecting on their errancy. Such "paradigm shifts" (which must be thought differently because they are without theoretical unity before or after their displacements) are often displacements of terrain brought about by a process such as this, which reconfigures a field almost behind its back, or beneath the level of its conscious aims. Moreover, I should emphasize that to whatever extent I have sounded a critical tone in response to certain of Venter's interpretations, I would never try to argue that nothing has happened here. Deconstruction could never be a simple nihilism or skepticism, as if we were to simply throw the vast apparatus of science and technology into a hole, but could only be *what happens* in and as this technoscience, whose most profound transformations perhaps always take place *sub rosa*, not just as we think or say.

The very structure of patent law is exposed to such deconstructibility. The operation of the legal system betrays symptoms or effects of deconstruction precisely by attempting to circumscribe and control them. With more space, I would try to show how everything I have demonstrated here about the coinfection of natural science and legal casuistry coheres with Derrida's deconstruction of the philosophical underpinnings of technoscientific invention and intellectual property in his "Psyche: Invention of the Other," where he writes, "These distinctions [between author's right and patent, between scientific idea and industrial exploitation] are not just hard to put into practice (hence they spawn a very refined casuistry); they draw their authority from 'philosophemes' that have in general received little criticism" ("Psyche" 37; cf. Derrida, "Des Tours" 127). A deconstruction of systems of metaphysics pulls the rug out from under juridical authority, while a deconstruction of systems of law reveals metaphysics to already be exposed to the contextuality it would aspire to exclude. We can observe this in the example of two concepts whose typical function we have already seen in response to Venter's patentmania. These are the concepts of nature and abstract idea: one can patent a "process, machine, manufacture, or composition of matter," but not an abstract idea, or a law, phenomenon, or product of nature (USPTO). Both of these distinctions, between abstract idea and concrete invention, nature and *technē*, are ways of attempting to control iterability. They distinguish a derivative iteration, an application with secondary status, from an original font of productivity, reproducibility, and self-identity (such that Promethean man can seek patents, but animals or gods cannot). Nonetheless, we have already seen that any specific product (Venter's products were just examples of this law, of iterability) is nothing without a supplement that produces inevitable effects of abstraction — which is to say that originary possibility is no easier to pin down than derivative actuality. What this process, machine, or composition here is, what it can accomplish and what contexts it can be set to work within, depends on an interpretation or reading of it that is produced and codified by the patent system, rather than being a bare fact from which the courts can derive their authority and decisions. Is Venter's invention a process for transplanting any genome into any cell, or this kind of genome into this kind of cell — and what is that, a kind or species? No matter how general or restricted our interpretation of it is, the effects of "abstraction" remain. Similarly, the idea of an originary possibility and productivity in relation to which the inventor's labors are derivative reconfigurations and manipulations, applications of a principle to a case, shapes the nature/*technē* distinction. It is not at all self-evident whether Venter's minimal genome is natural or artifactual — just as any "product of nature" is the result of conceptual and technological tinkering, ingenuity, and invention — most fundamentally because nature "itself" is artificial, what we sometimes call a *legal fiction*. "Natural law," which would be defined by its very inevitability, the automaticity of a matter and life that needs no mechanisms of oversight or legislation, a repository of immunity from the quibbling of courts and legislatures and the heteronomy of positive law and its respect or enforcement, requires the supplement of positive law for its recognition. It is a matter for the courts what will have counted as natural law. There is nothing more natural than this artifice or legalism, which challenges Venter's pretensions of control as much as it does those critics who are certain they know that something natural is here being misappropriated.

Given this undecidability, it is not accidental that the law works for and against its own purposes at once. Patent law justifies its existence by promising to incentivize innovation, but at every turn it must inoculate or autoimmunize itself against the risk that its very operations will hinder innovation. In theory, at least, it aims to reward invention without creating claims of ownership so broad that no further progress or novelty is possible. Even if we see this incentivization argument as the ideology of a ruthless capitalist machine that exists only to expropriate for the sake of the wealthiest, we should still recognize that it cannot be an accident of any economic, political, or legal system that the law works for and against invention at once. Venter has been compared to Bill Gates as a figure attempting to gain just this type of control over a still fledgling industry, amassing early patents over emerging technology that will give him a stake in all future profits. Industry watchdog (and occasional poet) ETC Group has referred to this venture, this Venter capital, as "Microbesoft" (qtd. in Calvert 173).[26] I will note here simply that this autoimmunity of the positive law follows inevitably from iterability — it is not a sign that we have failed to configure the economy or its legal bulwarks correctly (transformations of economy, polity, or law, could change this area fundamentally, but it would not be a change from iterability-undecidability-autoimmunity to their opposites or their sublation, as if there were such a thing). We will always be left to negotiate between natures and abstractions.

Who or what invents? This iterability is at once the greatest risk and resource for every claim of paternity, of having fathered the future. One symptom of this insecurity (and of the ruses of appropriation it makes possible) is legible in effects of signature, of how a proper name attaches and detaches, like an eager parasite, to scenes of authorship or invention. (I have just argued that "nature" is itself a kind of invention, and we see these signature-effects in the significant efforts devoted today to attributing the invention of nature to the name of "Descartes," a strategy with risks and rewards for every party involved.) Similarly, we have seen that Venter would like to be recognized as a great inventor, of the present and the future, not only for the commercial possibilities attached to such recognition (after all, he may not even be with us long enough to profit from these patents, though they may help ensure the future of the institute that bears his name), but also for a broader form of recompense, perhaps that of being seen as a scientific genius. Genius would be the capacity for invention. In the same essay I mentioned above, "Psyche," Derrida observes the necessary reasons why the concept of invention has an insecure relationship (perhaps we could say an insecure attachment style) with the signature or proper name:

> For there to be invention, the condition of a certain generality must be met, and the production of a certain objective ideality (or ideal objectivity) must occasion recurrent operations, hence a utilizable apparatus. Whereas the act of invention can take place only once, the invented artifact must be essentially repeatable, transmissible, and transposable. Therefore, the "one time" or the "a first time" of the act of invention finds itself divided or multiplied in itself, in order to have given rise and put in place [*d'avoir donné lieu à*] an iterability. The two extreme types of invented things, the mechanical apparatus on the one hand, the fictional or poematic narrative on the other, imply both a first time and every time, the inaugural event and iterability. Once invented, if we can say that, invention is invented only if repetition, generality, common availability, and thus publicity are introduced or promised in the structure of the first time. [. . .] To invent is to produce iterability and the machine for reproduction and simulation, in an indefinite number of copies, utilizable outside the place of invention, available to multiple subjects in various contexts.
>
> [. . .]
>
> Universality is also ideal objectivity, thus unlimited recurrence. This recurrence lodged in the unique occurrence of invention is what blurs, as it were, the signature of inventors. The name of an individual or of a singular empirical entity cannot be associated with it except in an inessential, extrinsic, accidental way. (34, 36)

First, Derrida explains an aporia or double bind within the concept of invention, interweaving invention and convention. Invention implies the notion of a first time, an unprecedented discovery, and yet it requires the creation of something whose status can be identified according to existing theoretical and legal conceptuality and can be put to work outside the immediate context of its inaugural moment. It is not invention unless it produces a kind of machine for re-instantiation or citation, and thus the very notion of invention's "first time" is compromised in advance, depending on its future appropriations for its status. For this very reason — because any "invention," whether we would call it a machine in the technological sense, operates like a machine in that it dispenses with the connection to the moment of its institution and institutor, functioning in the absence of author and authority — invention has a fragile relationship to the proper name (Bennington, "Aberrations"). Countless syntheses of life and the lifelike are sure to follow Venter's "minimal genome," but just who invented their possibility, in whose name they are created, will inevitably require the active intervention of a reading (and often, of a legal dispute) to determine. It rests on a trace-structure. It is not just the name of Venter that stakes its claims here, but Leduc, Loeb, Langton, many past and to come who might be sought out for their parentage and patentage. It will never be self-evident who invented that future or to what proper name it returns, what it means to be an inventor, or to be *in Venter*.

This impropriety of the proper name, one's own lack of ownership over it, may always give rise to insatiable manias for ownership, self-branding, and self-promotion, to the appetites of the hungry ghost.

It follows from the aporia or double bind of invention-convention that invention is never the object of a positive knowledge. In fact, the greatest risk to all inventiveness inevitably takes the form of the effort to program or codify invention. Any attempt to create a system, whether conceptual, legal, financial, or any other, that can anticipate and recognize inventiveness, precisely because it can only recognize what conforms to its conventionalities, will jeopardize those inventive ruptures that carry away or sneak past the guards and guardrails even of our concept of invention and of everything we thought we could anticipate under that name. Derrida refers to this programmation of invention as the "invention of the same," which sustains a non-oppositional relationship with the invention of the other ("Psyche" 39). The invention of the other cannot be specified by a separate set of conceptual or legal rules but rather is what displaces any such system without belonging anywhere within or without it. If this neutralization-by-recognition makes invention impossible, that is its only resource — the only true invention would be an impossible invention, one unrecognizable within any of the confines of present knowledge and power. Venter's constant drive to declare himself the first is an example of this programmation of invention, as is the operation of any patent office. Importantly, that does not mean that they do not invent or deal with inventions, but only that whatever inventiveness crosses their paths is not precisely what they think or say it to be.

Academic writing on synthetic biology is prone to its own form of this programmatic invention-of-the-same, and in this it is representative of much broader tendencies in academic thought today. Particularly among sociologists and anthropologists, one finds a series of publications eager to cast synthetic biology as a novelty of historic import (always with the concomitant boon of making the theorist the first to declare this novelty). Beginning with Nikolas Rose's *The Politics of Life Itself,* and including works by Melinda Cooper, Kaushik Sunder Rajan, Stefan Helmreich, and Sheila Jasanoff, a way of framing synthetic biology emerges that, even when it sounds a critical or alarmist tone, accepts the basic premises found in the self-descriptions of Venter and many other biologists.[27] In particular, it is said we have crossed into a new "age," "epoch," or "era" in our technoscientific control and economization of life, because the synthetic biologist now edits, engineers, invents, and markets "*life itself.*" Frequently, this apocalyptic warning invokes a Foucauldian conceptuality, by trying to fix the historico-conceptual limits of a new episteme, one that would be an inflection point in the evolution of "biopolitics." Even or especially when this millenarianism is invoked to denounce abuses of power or declare the need for oversight, regulation, or ethical restraint, it has already accepted the basic representation of life that we have found both undergirding and undermining Venter's most self-assured "inventions" (for example, equating life with its genetic code). What is that, "life itself?" It is not irrelevant to note that the question seems most elusive right here, that knowledge and even a certain power over life most evades those who make it their life's work. The science of biology is not where life is defined, but where it is placed in question. Life has always been the negotiation and suspension of its apparent borders, the exposure to every risk and boon of virality, of an other of life without which anything worthy of this name would be impossible. It has been, since before its beginning, synthetic life. This non-self-identity works within the self-certainty of any inventor, scientist, or social scientist who grants themselves the knowledge or power to speak for "life itself."

In no way does such deconstruction council some sort of complacency, an indifference or indecision in the face of happenings that are undoubtedly of momentous import for the future of what we call life. Rather, the risk of a simple conceptual circumscription of the present (a distinction between genetic engineering and synthetic biology, natural and man-made life, control over the epiphenomena of life and life "itself," or between the Fordist productive economy and post-Fordist financialization, the latter speculating on life "itself") is that we suppress certain demands and resources of resistance by oversimplifying the appearance or specter of what threatens us. All too briefly, I'll illustrate this by drawing on the brilliant work of Melinda Cooper, which merits a more thorough dialogue I hope to return to elsewhere. Cooper's *Life as Surplus* sees as a definitive trait of the present the *financialization* of "life itself," turning the life sciences into an investment vehicle within a post-Fordist economy with a novel relationship to debt and speculation. It would not be possible to exhaustively analyze this concept here, let alone to settle intractable debates surrounding "neoliberalism" or other markers of the present. But everything I have introduced in this chapter could be understood as a symptom of such financialization, how certain whims of the market dictate what happens within the life sciences and turn every scientific labor into an at least potential investment pitch or investment vehicle. The historical and theoretical shifts that define this present, as Cooper describes it, include the transformation of the engine of economic growth from growing productivity to growing debt, and the concomitant expansion of markets of speculation and speculations upon speculations that exposes all valuation to the cycles or bubbles of hype machines.

Now, if we were to place in question or perhaps deconstruct certain premises of this epochal distinction, it certainly would not be for the sake of denying the importance of what is happening today. If we were to point out, for example, that there has never been a life without speculation, that is not to be naïve about how certain financial institutions have harnessed these forces today, or about the unprecedented risks this economic system poses to the very future of life on earth. It is simply to say that we may misrecognize or misdiagnose those forces we most hope to oppose if we conceive of our resistance in terms of an opposition between speculation and non-speculation — as the return to a once and future non-speculative and non-spectral life. Moreover, when we read that neoliberalism "profoundly reconfigures the relationship between debt and life" (Cooper 10), that it divests itself of all "institutional reserve or foundational value" to project a "speculative future" without national foundation (10), that it inaugurates an "essentially promissory market" (28), no longer driven by "rational expectations" but by "collective belief, faith, and apprehension" (10), that we languish under the rule of "a world empire that is curiously devoid of tangible reserves or collateral, an empire that sustains itself rather as the evanescent focal point of a perpetually renewed debt and whose interests lie in the continuous reproduction of the promise" (30), that even this world and its "terrestrial reserves" may not suffice for a "debt form" aspiring to "return to the earth, recapturing the reproduction of life itself within the promissory accumulation of the debt form, so that the renewal of debt coincides with the regeneration of life on earth — and beyond" (31), and that "*life, like contemporary debt production, needs to be understood as a process of continuous autopoiesis, a self-engendering of life from life, without conceivable beginning or end*" (38; original emphasis), there is more than one indication that, if we still insist on thinking of these effects as the operations of an episteme or paradigm, it did not simply begin in the 1970s. Cooper will not have been the first to call for a life forgiven of its debts. If we are bearing witness to the descent to earth of a worldwide community of faith, dissolving national borders and "terrestrial" investments or talents for the sake of a beyond, a sacrifice of immediate or self-identical life for the sake of speculating on rewards waiting in a cosmopolitan kingdom come, of "the wholesale capitalization of a surplus life to come" (Cooper 49), we could be excused for feeling (somewhat like Philip K. Dick) trapped in time at the dawning of a Christian era. It is not some mere idealism or an ahistorical obscurantism to recognize these continuities, and in fact they are relevant precisely because they may hone our perceptions of the strategies necessary to intervene in the present. Specifically, if it cannot be the case that just yesterday or a generation or two ago life *became* indebted, or speculative, and if our generation will not have been the first to call for the absolution of original sin, then whatever we are capable of collectively bringing about tomorrow, even if it were to succeed in transforming the greatest and most haunting threats in the present, will not be a transformation from speculation to non-speculation. It would be a matter of no longer thinking of debt as a fall from an original plenitude, but rather as the nature of life or the nature of nature, without need of redemption or redeemer. Our present or our recent past will not have been "the first political form [. . .] to inscribe its relations of debt at the level of the biological" (Cooper 8), but it is precisely starting from here, from the synthetic origin of life death, that we are exposed to all chance and risk by a speculation that is the very opening of a future.

Computer Forbears

The previous section began by introducing Venter's arguments that the projects of perfect engineering control of life (including its commercial applications) and of perfect theoretical knowledge were of a piece. Minimization was supposed to accomplish both at once, by revealing the basic parts list and correlated functions of the essence of life, and thus making life a fully predictable machine. Or vice versa.

The aporias brought to the fore by this representation intersect or interweave with the multiple framings within which Venter always justifies his research. First, the engineering ideal and our practical aims may well be conflictual — rather than fine-tuning or improving the function of the cell, our prefabricated notions of efficiency often seem to hinder it (a "simpler," "rational" genome or cell may be slower, less robust, etc.). Thus, second, the grand-sounding knowledge claims of Venter's projects and their technological promise (both of which are inextricable from the work's reception) may well be in conflict; if a minimal cell ultimately offers a preferable chassis for bioengineering, it will most likely be because it has *reduced* or cordoned off the "complexity" of life rather than having *understood* it. Our genetic circuits and metabolic engineering may become more predictable and calculable only because we have recreated an imitation of life that obeys our commands, not by achieving perfect knowledge but by excising whatever stubbornly resisted our will.

(The invocation of *complexity* whenever the materials of the bioengineer prove unpredictable presupposes a sort of ideology of science within which even the unknown is predigested. If the operation of a genetic circuit, for example, is not perfectly predictable, attributing this to "complexity" implies that its parts interact with the other parts of the genome or cell with nonlinear results. Thus, it assumes a mechanistic and deterministic system and implies a telos of perfect intelligibility for what is not yet well captured by our models.)

It is not that knowledge and power are simply in conflict with each other, but that they are in conflict with themselves — they are necessarily exposed to effects of contextuality. There is knowledge or power only within something that functions like a context, and thus they remain exposed to a threat or promise of expropriation that is their condition of possibility, their "essence." This suggests a different reading of a famous aphorism that has become a sort of motto of synthetic biology, "What I cannot create, I do not understand." This phrase was found written on Richard Feynman's blackboard after his death, and Venter encoded it into the watermark he added to JCVI-syn3.0 (using DNA base triplets as a cipher for the English alphabet) (Venter, *Life* 124–25). Evelyn Fox Keller has called the near-ubiquitous invocation of this mantra among synthetic biologists "ironic," and reserves judgment regarding what Feynman could have meant by it (Keller, "What" 295). The synthetic biologists clearly deploy it as the apotheosis of their engineering ideal — a life they can completely control is one they think they will have understood. We have already seen that their modus operandi is closer to "what I can create, I needn't understand." Indeed, some practitioners state this explicitly, such as Tom Knight: "an alternative to understanding complexity is to get rid of it" (qtd. in Various). On the other hand, Feynman was an advocate of the contemplative ideal of knowledge for its own sake, exemplified by the rational systematicity of theoretical physics. Without pretending that I know any better than Keller what those words could have meant to Feynman, I will suggest a possible reading simply to elucidate the constitutive undecidability here. It would be possible to understand these words attributed to Feynman as giving expression to the rational form of scientific or theoretical thought — nothing has achieved this form unless it is deducible from first principles. If one has this form of understanding, one can write a stable synchronic set of mathematical equations, and the entire history, for example of the universe since the Big Bang, will unfold before our eyes. In practice, this takes the form of simulation or modeling, but one could understand what Feynman was calling "creation" in this way, as the ability to re-create the given from first principles. If the engineering ideal of the synthetic biologists appears at once to be as different as possible from this, as different as practice from theory, and yet to be synonymous and indistinguishable, it may be because of the difference that inhabits the inside of both creation and understanding, knowing and making, or knowledge and power.

If both "theory" and "practice" invoke rational design from first principles, it is because what both most have in common is what they resist facing in themselves. The origin or essence we hope to reclaim never was, and thus remains interminably to come. Even if one can only speak *as if* one had recaptured such a force, one only ever does so *starting from here*, from an ungrounded drift among terms and germs whose derivativeness is originary. Neither "knowledge" nor "power" exist in any purity, but rather this opposition and every nearly synonymous substitution takes form in a vain attempt to stabilize an undecidability that inhabits "both." Because neither ever arrives to or from an origin, there is only the simulation and dissimulation of creation or understanding, one toys with a momentum received from an anonymous alterity, by a kind of cunning or deceit, a *mēkhanē* more fundamental than any mechanism. The artificiality or syntheticness of everything descends from this necessity — it is not a product of the self-possessed creativity of the ideal engineer, but the inner difference that haunts the most vital appearances as their condition of possibility and impossibility. Life or nature appears as a machine because it appears as a model, as something that does not have its origin or impetus in itself but receives it from an elsewhere that has never finished arriving — the endlessly rediscovered mechanism that we have found recurring through the history of biology starts from here. Nothing has ever simply lived, nor does anything die once and for all, but life death is just this machinating putting-to-death or treating as dead so that the other than vital, sur-vivance, might have its chance or risk.

What could such a life carry or bear, what are its transferences and metaphors, what monsters does it ready? Venter thinks he recognizes the descent or genealogy of Synthia, and what he sees is not, is never, pure misrecognition. He describes his experiment as "the first living self-replicating species to have a computer as a parent" (Venter, *Life* 125). It is surely only habitual, but not at all uninteresting, that he elsewhere pluralizes this parentage, suggesting a sexual difference within the machine: "the first species . . . to have its *parents* be a computer" ("How Scientists"; emphasis added). If there is misprision here, it is only this "first"; it is never the first. All life descends from something like a modeling, an artifice or synthesis no more living than dead. We all have some computer in our family tree. It was rightly pointed out to Venter that he had not simply created a "new species" or "new life," nor had he designed his synthetic cell from "first principles," because it was still largely dependent on a "natural" model, an existing cell (Bedau et al.; Dan-Cohen 29–30). There are only such imitations of life; "natural" life receives its first chance by imitating itself, by mimesis (nature, Aristotle says, is what happens *always or for the most part*), and thus the difference of modeling or the synthetic is *within* the "natural." Nor should we be so quick to assume that these synthetic cells are the first and only to merit the nickname *M. laboratorium*, suggesting its birthplace is the laboratory.[28] It is not exactly a contradiction that Venter declared the computer to have parental status over this child, and that he clearly took pride of parentage in it himself. He describes the scene of JCVI-syn1.0's birth with all the trappings of a proud father: he mentions arriving in the lab that day with a bottle of champagne and a video camera, and one of his team members describes placing a happy birthday balloon by its incubator (Venter, *Life* 123).[29] The first organism born of computer would be the first born of man, (no longer dependent on the other, which means nature as woman or woman as nature, evidently), according to an implacable logic of which this will certainly not be history's last example.

One has the feeling, in all this, of a striving to overcome a certain guilt or lack, a finitude. Perhaps what motivates Venter, and so many others, is an *original syn*.[30] He even referred to his minimal genome project, on certain occasions, as "the Genesis Project."[31] Their ultimate aim would be to create life by means of a fully autonomous production, both in that it had full knowledge and self-knowledge of the secrets of life and death, and in that it would not depend on a sexual or reproductive partner, or any organic form or model that was pregiven. The press may have been communicating this aspiration faithfully when, despite the many women involved in this research (including the first author of several of their most important studies, Carole Lartigue), it relied on monosexual language to communicate the feat: "man(-)made life." Ultimately, the quest for simplicity has only displaced and reconfigured its necessary dependence on something received from elsewhere, from an alterity that can never be commandeered within the full light of presence. This *syn* is not the union of the dual, but an originary difference that makes the values of the technic or machinic — heteronomy and virtuality — more fundamental than any nature. If there is a lingering religiosity in the dream of a life that would be purely known and made from first principles, by an ontotheological unmoved mover, an original syn is neither a curse nor affliction upon this vision, nor is it the promise of apocalyptic redemption. It is the gift of a *syn* without fault, fall, failure, lack, or guilt — without pure life or knowledge before it or beyond it. Even its finitude is undecidable — any limit it attempts to place between itself and the other is only a substitutive projection, like a theatre's curtain, an inner and artificial difference that finds us on both sides without identity. Syn should not be mistaken for an essence (or a minimal or universal structure, anything that could repeat or reproduce itself), for the same reason it cannot be thought as a parasite of some essence that is there already; what gives anything its first chance or risk makes anything like a simple origin (of life or anything else) impossible — it leaves every essence as simulated or framed, haunted or machinic, as what cannot run on its own or of its own, possessed without self-possession, neither living nor dead, exposed to an over-taking that belongs no more to self or other, neither recuperation, restoration, self-realization, re-iterating without identity, derivative without origin, sur-viving without end.

Chapter 5

"De-Extinction"

The Specious Concept

I come to the conclusion that the period of gestation covered eighteen months. This period of exactly eighteen months might suggest, at least to Buddhists, that I am in reality a female elephant.

— Nietzsche, *Ecce Homo*

Elephants are contagious.

— Paul Éluard

In the previous chapter, the instability of the concept of species was shown to interfere with synthetic biologists' attempts to gain knowledge and power over life. In this chapter, a similar pattern will be elucidated within the domain of de-extinction research, affecting certain conservation projects that hope to preserve life from the risks posed by climate change.

Like the "minimal genome" or "synthetic biology," the name de-extinction *has become a galvanizing force in scientific research and the popular imagination — without naming any one thing in particular. It has attracted the most attention by claiming it will be able to "resurrect" extinct species including the woolly mammoth. In practice, this new name covers a range of potential approaches to conservation work, some of which draw on novel technologies,* while others are quite old practices merely described in a new way. For example, one can restore the traits of an extinct species simply by back-breeding their living relatives, meaning that one uses an age-old intervention, artificial selection, to breed individuals with the desired traits.

The novel technologies used to justify the strongest claims of "de-extinction" are adapted from synthetic biology. Projects such as woolly mammoth "resurrection" plan to take ancient DNA sequences recovered from frozen mammoth remains, synthesize them, and splice them into the genome of a cell from the mammoth's closest living relative, the Asian elephant. They will then use a process called somatic cell nuclear transfer, often referred to as "cloning," to insert this edited genome (a hybrid of elephant and synthesized mammoth DNA) into an elephant embryo. The aspiration is that these genes will govern traits such as long hair that allowed for the appearance and lifestyle of the mammoth, and thus that these "mammoths" could be relocated to Siberia where, some argue, they will restore vital ecological functions.

Elephant reproduction is normally sexual, which means that cells from two partners combine to form an embryo. An elephant's somatic cells, all the cells in their body except the germ cells, are diploid, meaning that they contain two copies of each chromosome. The germ cells, sperm and eggs, are haploid, containing just a single copy. An embryo formed through sex combines half of each of its parents' DNA to form an embryo genetically distinct from both its parents.

Somatic cell nuclear transfer is popularly called cloning because it creates an individual that is genetically identical to its parent. It does so by removing the genome from an egg cell and fusing that enucleated cell with a somatic cell containing the parent's genome (or by transferring just the nucleus). In the case of "de-extinction," this somatic cell would have been edited to include some genes from an ancestor species. Still, because it relies on the same technology, the process is often referred to as "cloning" a "mammoth."

"De-extinction" projects raise two intimately related questions. One, relatively theoretical: is an organism created by this process really a "mammoth?" The other, relatively practical: does such a project really further the goals of conservation — is this what we should be trying to conserve? These questions have given rise to intractable debates among all those involved in de-extinction and conservation work today. In this chapter, I demonstrate that aspects of these debates follow from the instabilities of the basic concepts or categories that are at stake. It is impossible in principle to arrive at a rigorous definition of species, or of a particular species such as the "mammoth," and this impossibility overdetermines scientific attempts to gain control over life and its reproduction (for example, the attempt to "resurrect" a "species"), as well as our efforts to protect and conserve existing species. The deconstructibility of basic categories means that our technological tools and ethical commitments shape theoretico-scientific concepts, while questions of an abstract, philosophical form (such as "what is a species?") form and displace the ethical demands confronting us in the present.

***

The genome-editing technologies described in the previous chapter have mobilized the energies of certain researchers who feel they can be used to revolutionize the work of ecological conservation. It is claimed that we could bring back organisms or traits that have been lost to extinction, and despite the many asterisks that the name requires, this project is widely known as *de-extinction*. Those who have been most effective at attracting a spotlight to this project claim that it will be possible to "resurrect" long-lost species such as the woolly mammoth or passenger pigeon, whether to restore ecological balance or expunge a guilt incurred by our species. As with "synthetic biology," the claim to have designer-control over the living and the dead, and even the claim of technoscientific novelty that gathers around the name *de-extinction* can be placed in question. Some practitioners promise to "clone" extinct organisms by redesigning their living relatives' embryos to include lost genetic sequences; while there are undoubtedly new technological capabilities that make this even possibly possible, it is noteworthy that other practitioners of "de-extinction" hope to achieve similar effects through a much older process: *back-breeding*. Not only might this achieve the desired effect at a phenotypic level (producing look-alikes of extinct organisms), but it is also no less a manipulation of the mechanisms of heredity, including the genes. Artificial selection, one of the oldest human technologies, and one that is not without pre-human precursors, is in its own way a genetic engineering. In the 1980s, Richard Lewontin and other dialectical biologists could already make fun of their generation of genetic engineers for bragging about accomplishing in the lab what could be done more easily in a greenhouse (Levins and Lewontin 223–24). Once again, my intention in recognizing these haunting perseverances, or the misrecognitions of novelty and power, is not to claim that nothing ever changes, but that it is always on the reverse side of vision or consciousness that the unanticipatable seizes us.

As we have recognized before, it is not a simple task to separate the "theoretical" and "practical" preoccupations attending de-extinction, or to distinguish its "reality" from "rhetoric." One can still make such distinctions, and in a sense one must, but they take place within a structure or frame that can itself be placed in question, and thus is no more theoretical than practical, no more literal than rhetorical, in that all such values descend from this non-saturable contextuality. With respect to the present case, questions about how best to utilize limited resources and efforts of conservation, about appropriate responsibility for ecological stewardship, about the urgent demands conjured by imminent climate catastrophe, are not purely separable from a question that has long vexed biologists and perhaps all the living: what is a species? Just as synthetic biology's quest for the minimal and essential seemed confused as to whether it was answering the question "what is life," or doing away with it to make a calculable engineering material — or whether it could do one without the other.

As always, these questions are conjured and displaced by ingenuities of technological reproducibility, even if we cannot be sure without such interpretations that we wield a novel power. In this case, the claim that we might "clone" a mammoth (for example) draws on a technology similar to the one that "cloned" the famous sheep Dolly, a process less colloquially known as somatic cell nuclear transfer (SCNT). It is now a commercially available technology, often sought out to replace lost pets. A polo team even recently won an Argentine tournament with several "cloned" horses (J. Cohen; Malin et al.). Nonetheless, already in 2000, it was possible to place in question the pretense implicit in the name *cloning*, that we are reproducing not just species-identity but individual identity. Evelyn Fox Keller's *Century of the Gene*, a study galvanized in part by the success and failure of the Human Genome Project (by succeeding in sequencing our "genes," or representative examples of them, it failed to vindicate our models of genetic agency), explores the many inadequacies of the model of heredity as a "genetic program." She also critiques how this model informs the project of "cloning" (Keller, *Century* 87–93). Like humans, a sheep's somatic cells are diploid, containing two sets of chromosomes, which consist of long strands of DNA tightly wound around globular histone proteins (these structures are more complicated than the chromosomes or genophores of prokaryotes — the field of synthetic genomics is nowhere near producing a eukaryotic chromosome). These diploid cells make up everything in the sheep's body except its germ line, the sperm and egg cells that are haploid, containing only one set of chromosomes. In SCNT, instead of uniting egg and sperm to obtain a diploid zygote, the haploid nucleus of an egg cell is removed, and replaced with an electrofused somatic cell or its nucleus.[1] SCNT allows for the creation of an embryo that is genetically identical to its parent (so long as we disregard the chimerism and DNA damage that is a normal part of life). Keller pointed to the following gaps in the interpretation of this genetic identity as "cloning": why, if these genes alone are supposed to determine the developmental potential of an organism, is it necessary to place them in an egg cell for "reprogramming," in order to create something that can proceed along the course of embryonic development (which will also require implantation into a womb)?[2] What do the parts of the donor cell beyond the genetic sequence, and the parts of the egg cell that receives it, and the organism that gestates it, contribute to development, to life? Keller pointed out that this paradigm is part of a long history (much longer than modern biology) of disregarding certain contributions of the mother to the form and formation of the child — this must be read together with the fact that the first cloners took their somatic donor cells from mammary epithelial tissue, and named the lamb Dolly after Dolly Parton, a woman they thought of as famous for her breasts (Hirsch; Weintraub; Callaway, "Dolly").

(Similar to the paternal pride of Craig Venter, and recurring in every instance of the de- and re-naturing of virality, a sort of artificial or spiritual family restores and recuperates itself around the strange graftings of these family trees.

The polo player who cloned his horses bred them through a business he owns, whose chief scientist is Adrián Mutto: "Zipping past verdant fields filled with pregnant broodmares, Mutto reflects, 'When I see a clone of mine born, it's this unbelievable sensation. I think, I've known you since you were one cell.' He parks the cart in front of the Hanna Montana clone's paddock and leans over the fence. Utterly uninterested in the scientist's arrival, the leggy one-year-old continues grazing. Trying to persuade her to come over, Mutto pleads jokingly, 'But I am your father!'" [H. Cohen]. As with any family, such paternal or parental display can always hide sinister self-concern. "Cloning" remains a poorly understood process, and thus one with a significant failure rate. It often requires one hundred embryos to bring one to term, and while many fail to implant in their surrogates, still others miscarry in the process of gestation or die young, both of which can cause significant suffering for the animals involved. Crestview, the business with which these men are associated, almost certainly lies about the health of its foals — it claims to have ten times the success rate of any other lab practicing somatic cell nuclear transfer [Malin et al.].)

Keller is also quite critical of the genetic determinism informing *Jurassic Park*, which imagined cloning dinosaurs by a process almost identical to the one being promised today for "de-extinction." (One plan for the translocation of a new "mammoth" population, should the project ever reach that stage, would place them on a nature reserve in Siberia maintained by a Russian scientist, a property he is attempting to "rewild" and that he calls Pleistocene Park.)[3] One salient difference is age: around the time Crichton's book was published, several teams of ancient DNA researchers claimed to be sequencing DNA from insects trapped in amber, in one case from a weevil that died 120 million years ago (Shapiro 53–57). These claims have since been debunked, and it now seems impossible to recover such ancient DNA (amber is not preservative at all) — the oldest DNA that has been sequenced at present is from about 1.6 million years ago, and it is necessary for organisms to have remained frozen from shortly after their death to recover even fragments of DNA from such lengths of time (Callaway, "Million").[4] While there are many such frozen mammoths, one can still only recover snippets of DNA information, even from remains so well preserved that flesh and hair may be intact. Also, there is necessary interpretation to the deciphering of ancient DNA — any sample will have been exposed to bacteria, fungus, and plant roots, in some cases to urine from grazing animals, and even the most carefully collected will have contact with human researchers — DNA from all these sources will be mixed with anything that can be sequenced from a sample, and the text one seeks is so fragmentary that one must in a sense know already what one is looking for to find it (one must expect, for example, that sequences that can be mapped onto an elephant genome, despite certain differences, will belong to the mammoth).[5] If one hopes, then, to "clone" a mammoth, there are certain technical questions or stumbling blocks: can one glean sufficient ancient DNA sequence information, can one edit the genome of its nearest living relative — the Asian elephant — to resemble it, can such a mammoth-like nucleus form a viable embryo in an elephant egg, can it develop to term in an elephant womb, how will these babies be reared and relocated so that they can survive in the wild, and why would we do all this when elephants themselves are endangered species? These practical-technical questions are not purely delimitable from the more theoretical-sounding question: what is a mammoth? Is an edited genome from what we call an elephant, implanted in an elephant egg cell and brought to term in an elephant womb or perhaps an artificial substitute (a technology that has yet to materialize), then reared and translocated by humans — is that a mammoth? Is it possible to justify this project or evaluate its success without broaching that question?

Debate over this species question inevitably arises both among the de-extinction project's boosters and its detractors. While there are aspects of the species question's occurrence that are specific or symptomatic of the framing of this particular endeavor, its undecidability is shared in every subdiscipline of biology. One finds as many responses (not all of them *answers*) to this question as one finds biologists, including those who would hope to decide the question (in dozens of conflicting directions), those who admit perplexity or wonder in the face of it, and those who claim it is an unscientific question that biology does better to ignore (Wilkins). This pattern is symptomatic of basic or grounding "concepts," which produce all manner of overconfident resistance and mystical openness — *what is life?* has similar effects and is almost a substitutable query. It certainly is not a question that could be answered "empirically" (in fact we are confronting here the internal limit of all empiricism, its presupposition of decisions of a rationalist, theoretical, or ideological type) — if we want to turn to the world and glean from it how the lines among species should be drawn, we can only do so if we already have some sense of what we are looking for, on the basis of an at least implicit decision as to what counts as species and the divisions between them. The necessary arbitrariness of this decision is insurmountable. Moreover, nothing can absolutely promise us that *species* is a name for some one thing that simply waits to be defined. We will never surpass the limits that make this a word in a given language, a given culture (of science and of an English science, however globalized), within given formations of politics, race, sexuality and gender, and so on. On the one hand, all of this dissension over the term can only take place on the basis of an inheritance of precomprehension or background understanding that necessarily exceeds and makes possible-impossible scientific reflection upon it and precise (if conflicting) definitions. On the other hand, every means we have of naming the source of this precomprehension, attributing it to "culture," or to a particular language, sexual, racial, or political formations, family structures, and so on, descends from the "same" inheritance. None of these cultural designations are prior to categories that stand in for "nature," not because nature is prior to culture, but because "both" share the same belatedness or afterwardsness (their definitions are rationalizations). It is just as possible to recognize that each of these differences or divisions (those we name as sexual differences, racial differences, linguistic differences, etc.) even if they overdetermine a category such as "species," can nonetheless in turn be subdeterminations of it, can take shape as stand-ins of species difference. ("Gender," for example, is a species of species, "species" a gender or genre of gender, and we turn in this circle without the possibility of grounding or hierarchizing these substitutions.) In which case, there is no possibility of a purely theoretical discourse on species, of a metalanguage that would not itself fall under the classifications it pretended to determine. Instead of a theory or concept of species, there would be only a species of theory, or perhaps specious theory, specious concepts.

Rather than pretending that a practicing biologist can continue their work unencumbered by the question or questions of species, by casting it off onto linguists, historians, philosophers, literary theorists, or onto cultural studies or other disciplines, everyone including the biologist operates at once, even if unbeknownst to themselves, as a philosopher, a linguist, a historian, a sociologist, a poet, and so on, dependent on decisions that summon the specter of each of these fields or responsibilities without ever offering a resting point for the authority or identity that could satisfy them.[6] Moreover, even if one hopes to preserve the authenticity and self-sufficient authority of science by pretending that such disturbing questions belong elsewhere, it is inevitably the case that there is no way to act or speak as a scientist at all without invoking designations that function as species-difference does. One must name, one must attribute one's conclusions to this class of entity and not that one, and one does not think scientifically at all if one deals only with inimitable singularity — which may mean that nothing is purely non-scientific. Nor is it as simple a matter as asking whether "species" is a "natural kind" — whether it belongs to nature or is artificially imposed there by biologists — for there is no "nature" to speak of unless something innate or inborn, the essence of species or a species of essence, has been posited.

Species, like life, is sometimes defined with reference to *reproducibility*, as a reproductively isolated population. Rather, species is "defined" and de-limited by its *iterability,* by what makes repetition, identity, and everything else depend on or descend from a différance without end. There is no species without sur-vival, an openness on the to come that makes every definition or border provisional and artificial. There is an element that remains artificial, arbitrary, symbolic, cultured or cultivated, and machinic or heteronomous about every demarcation within the living, the decisions that mark a-filiations, territories, and matings or familiality among the living, a natural lection that is the very science *of* life.

Proxies: Mammoth, Mammary, Mammal, Mama

The domain where the call of species, or anything else, is heard is not where it is *answered* but where it is most *in question*. This is what one finds among the researchers of various disciplines who have been brought together in amity or enmity to weigh the possibility and ethics of "de-extinction." There are those who claim they will advance the demands of conservation work by creating technologies of "de-extinction," and those who argue the true task of conservation may even be undermined by it. Among the debates that shape these starkly opposed attitudes is a dissension around the meaning of "species": is an edited and cloned genome, gestated not just by a surrogate of another species or an artificial womb but also *ex situ* in some laboratory and later translocated to a new habitat, really the species we say it is? Which means, in other words, is this what we are trying to *conserve?* In 2014, the Species Survival Commission of the International Union for Conservation of Nature (IUCN), an agency that, among other responsibilities, maintains a listing of endangered, threatened, and vulnerable species, met to evaluate the questions raised by "de-extinction." IUCN began its report by questioning the definition of species that was implicit even in the name:

> The term "de-extinction" is misleading in its implication that extinct species, species for which no viable members remain, can be resurrected in their genetic, behavioural and physiological entirety. These guidelines proceed on the basis that none of the current pathways will result in a faithful replica of any extinct species, due to genetic, epigenetic, behavioural, physiological, and other differences. For the purposes of these guidelines the legitimate objective for the creation of a proxy of an extinct species is the production of a functional equivalent able to restore ecological functions or processes that might have been lost as a result of the extinction of the original species. Proxy is used here to mean a substitute that would represent in some sense (e.g., phenotypically, behaviourally, ecologically) another entity – the extinct form. Proxy is preferred to facsimile, which implies creation of an exact copy. (IUCN SSC 1)

IUCN makes at least two important claims here, the second of which I will return to in a moment. First, on the assumption that only a "faithful replica" can be said to belong to a species, it argues that "genetic, epigenetic, behavioural, physiological, and other differences" necessitate that we designate these re-creations as *proxy species* rather than their "de-extinct" originals. On one level, it is necessary to affirm this precaution; it is undoubtedly the case that the most headline-grabbing spokespeople for "de-extinction" have tried to profit from a technofuturism that pretends we have a full understanding of the nature of life and heredity, and thus only a few technical challenges remain to produce or re-produce it at whim. In fact, there are immense unknowns both as to what effects "genes" may have on development and as to what factors beyond the genes as traditionally understood — epigenetics, plasticity, social interaction, and more — actually bring about the traits associated with a species. It is necessary to emphasize these unknowns when an overconfident scientist claims they know the *logos* or formula for a "mammoth."

Yet, it has never been the case that something has needed to stay the same to count as a single species. To some extent, it has always been an arbitrary decision of the investigator what will count as a "mammoth" or anything else, which complicates the interpretation of these unknowns. It is precisely our power over this designation that thwarts us — the drive behind this work cannot be satisfied by "it's a mammoth if I say it is." Ben Jacob Novak,[7] a biologist working with Revive & Restore, an organization promising to revolutionize conservation work with tools from synthetic biology, advocates for a particular understanding of "de-extinction" that challenges the IUCN's definitions:

> Considering a cloned organism a proxy has been built on the argument that a cloned member of an extinct species is epigenetically different from its historic ancestors — an argument made solely in the vacuum of de-extinction discourse which, if we accept this argument, casts an alarming realization over all of conservation. If a cloned Bucardo is a proxy of a Bucardo, then every species epigenetically altered by human activities is now extinct and has been replaced with anthropogenic proxies. This means every recovery facilitated by translocation, captive breeding, habitat restoration and so forth, are not recoveries at all but have rendered species extinct and substituted them with new forms. (Novak 2)

If even an epigenetic alteration betrays species identity, then all conservation work is self-contradictory — one can "conserve" only a kind of zombie or body snatcher, a replicant that is made extinct even by being saved. Novak is not wrong to question the IUCN's equation of species-identity with "faithful replicas," but his proposed definitions are just as aporetic, and it is this mirroring impasse I hope to elicit.[8] Both parties are correct in arguing that the other invokes a debunkable definition of species — yet both are dependent on an identically unstable definition of their own.

Novak's critique of the IUCN's "faithful replica" definition is undeniably correct: everything we have ever counted as a species, whatever the extent of its disturbance by humans, has brooked differences that are not even limited to the "epigenetic," however that is defined. Moreover, why stop at just us? Beyond "human activity," every interaction of the living may be producing such "epigenetic" changes. I imagine Novak might well be frustrated, though, by the inevitable conclusion that we can call whatever we like a "mammoth," and neither a god nor nature will come to our rescue or condemnation, to speak the true names of the living. It is this frustration with our own power, a power indistinguishable from impotence — an *omnimpotence* — that I am trying to tarry with here. Deconstruction follows from the undecidability that is resisted in these counterformations (of the IUCN and Novak) both of which resist, in opposite yet harmonized ways, recognition of this weak force (what changes everything while leaving it entirely untouched).

The IUCN's argument denies that such a "clone" or re-creation should be identified with the species it is modeled upon. This necessarily implies that there are some known limits within which species identity is secure. Novak argues that if we really held to this presupposition, of an ideal closer to the natural as the telos of conservation, conservation work would be impossible. Conservation almost always works with organisms that have encountered humans, and frequently must itself manipulate a species or its environs (thus affecting its "epigenetics" and development). The work of conservation can only change what it saves or save some part of what was at the expense of something else. Nonetheless, if Novak has seized upon a sort of bad faith within the IUCN's definition, he is no less mired in it himself — it does not follow, simply because one recognizes a certain hypocrisy in the IUCN's willingness to admit certain differences within the species concept and not others, that the differences necessary for "de-extinction" to proceed are therefore admissible. Not because the truth of this concept is found in a third locus or text, but because that truth never arrives.

We are called to decide — on a question that has the ponderousness of idle speculation or theory, yet mobilizes the practical and ethical urgency of irreversible loss in the face of climate catastrophe — without any possibility of the authority arriving that could reduce the *undecidability* of this category. Deconstruction takes place as such a decision in the face of undecidability — there is only decision where undecidability makes the simple calculation or programmatic application of a norm impossible, and thus every decision is haunted by an impossibility or impotence that makes it belong to the other, an act of faith.

Now, both the IUCN and Novak reach a conclusion that seems, at first glance, to make their disagreements purely academic. In fact, if we were to take them at their word, it is not entirely clear why or whether they disagree at all. The second point made in the passage quoted above from the IUCN report, that "proxy species" can only functionally replace a lost species by filling their ecological niche, is Novak's position as well:

> A corrected definition of the mode of de-extinction outlined by the IUCN is this: de-extinction is the ecological replacement of an extinct species by means of purposefully adapting a living organism to serve the ecological function of the extinct species by altering phenotypes through means of various breeding techniques, including artificial selection, back-breeding and precise hybridization facilitated by genome editing. The goal of de-extinction is to restore vital ecological functions that sustain

> dynamic processes producing resilient ecosystems and increasing biodiversity and bioabundance. (Novak 5)

Both definitions emphasize an ultimate aim of conservation that is agnostic or indifferent to the question of species — it renders whatever could be counted as species substitutable to the extent that a proxy can perform their functional role within an ecological structure. Nonetheless, under the umbrella of "ecological replacement," Novak creates a taxonomy that carves out a novel place for de-extinction and that reasserts a value for species identity. In Novak's eyes, "de-extinction" should be distinguished from "assisted recovery," both of which can accomplish this task of ecological replacement. Assisted recovery implies that the species in question was never extinct, and that it is being replaced with individuals from its own population, part of a genetically continuous lineage with the population they are supplementing (for example, if individuals from that population are bred in captivity then re-released in the wild). Interestingly, Novak's definition of species continuity depends solely on genetic continuity, which convinces him that certain "clones" should be counted in this category of assisted recovery. If the clone is made from a cryopreserved tissue from a member of the lineage being replaced (as was the case with the attempted de-extinction of the bucardo, a species of ibex that went extinct in 2000), this implies continuity of genetic lineage to Novak, and therefore means that the species in question was *never really extinct*. It is brought back from a "torpid" state, in Novak's preferred terminology, but his conviction with respect to genetic identity allows artificial reproduction to displace the very conception of extinction, just as a related form of technological reproducibility changed the meaning of "extinction" for epidemiologists grappling with the categorization of synthetic viruses (see "Introduction: Principles"). Then, there would be "de-extinction," according to Novak's terminology, only where ecological replacement depended on the manipulation of another species in order to restore traits or functions from a lost genetic lineage — this manipulation may include now traditional tools of conservation work such as translocation or back-breeding, or it may involve the new synthetic biotechnologies that allow for genome editing. One can see that Novak is attempting to domesticate these seemingly novel practices bearing the name *de-extinction* — by categorizing them together with time-honored tools of the trade — in order to make them appear uncontroversial in the eyes of his colleagues and perhaps regulators. At the same time, the eminently pragmatic focus that emphasizes ecological function only preserves a continuing concern with the true nature of species continuity. Everything in the at-first-blush tidy system of classification Novak has constructed trembles in the face of the undecidabilities of species: "The lines between assisted recovery and de-extinction blur for species with highly variable, distinct populations such as the wolf" (7).

In the case of the minimal genome, we saw that when genetic agency was placed in question, researchers fell back on a functional notion of similarity:

even if we cannot attribute essentiality (necessity and universality) to specific genes, we can attach each gene to a "nearly universal function" and claim that this set of functions represents the essence of life (Hutchison et al., "Design" 10). In the same way, we can see two ostensibly opposed conservationist discourses agreeing that the guiding aim of "de-extinction" need not be a true resurrection of lost species but simply functional restoration by filling an abandoned niche. This represents a shift from genetics to structuralism, from identity as genealogical or essential continuity to identity as substitutive placement within a system of differences. But the undecidability of species is already an undecidability of genesis and structure — these arguments substitute for each other so readily precisely because neither can rigorously ground identity or difference. Just what is the niche that needs substitutive replacement or the function for which we seek equivalence by proxy? What were the sum of the contributions of this "species," what behaviors might it have practiced that went unobserved or perhaps would only have been revealed in unprecedented ecologies yet to come? Structure is no more secure than genesis if we start from the necessity that "species" is *what we do not know*. Or that "species" remains insufficient for our aims to the extent that it can only name something that belongs to our knowledge and within our impotent power, omnimpotent, regardless of how we define or conceive it. Just what is lost when a species is lost? And when is a species lost? Or rather, what is lost or perhaps gained at any and every moment that may not be recognizable to us as "species" but may be more essential for life or rather for sur-vival, for what exceeds anything that has heretofore known itself as life, an excess that has always been constitutive of what we have called the living? That this question will never ultimately be answerable does not absolve us of, but rather exacerbates the responsibilities that are called for by ecological stewardship, which is never merely restoration nor replacement. It is because neither structure nor genetics or genesis can define the limits of our responsibility that we are called to a necessary and impossible task.

The surest symptom of this call, the sign that it has been heard, even if only to be resisted, is the nearly ubiquitous *resurrection* of species-thinking despite its having been declared unnecessary or replaceable. In this sense, species is always already what comes back, what is revived, restored, *revenant*, or de-extinct. The face of the other, the gaze that calls us to *re-specere,* respect, is always a species of specter or specter of species, a re-specter. We have already seen how the IUCN and Novak, even while arguing that ecological replacement was the true aim of conservation practice, made their ideas of replacement depend in certain respects on their divergent species-concepts. This movement of resistance and insistence, toward and away in a single gesture, rejecting and reclaiming, diverts itself from the site where its authority is manifestly unstable, by claiming it has no need to go there, while granting itself all the same values in another locale.

Beth Shapiro, whose lab sequences ancient DNA and who is collaborating on "mammoth de-extinction," exemplifies this pattern. In the most reasonable fashion, in her admirably accessible writing, she details all of the reasons that our knowledge and technology cannot really promise the de-extinction of a true "woolly mammoth," and invokes the functionalist justification of the project to give it rational and ethical appeal: "This — the resurrection of ecological interactions — is, in my mind, the real value of de-extinction technology" (Shapiro 10). It is said that placing hairy elephants adapted to cold weather in Siberia will create opportunities for increased biodiversity and will protect carbon that is trapped in the permafrost. Nonetheless, she calls her book on the subject *How to Clone a Mammoth*. Of course, she wants to sell books, but this is precisely my question: What is the appeal, the lure and allure, of this species name? Why does it attract our attention, not just among non-expert readers, but among the many scientists who have gravitated to this field and are often less cautious than Shapiro in claiming a power over the name? In other words, *why are we doing this?* What role does the promise and the secret of the name play in drawing us on like somnambulists?

In addition to this ecological-function argument, there are two substitutable species-specific arguments put forward in defense of this particular project of "de-extinction." Of course, it is often said that it will "resurrect" the woolly mammoth, but with no less frequency, and often within the same text, it is argued that it will protect the endangered Asian elephant by creating an elephant strain adapted to the cold weather of the less rapidly disappearing Siberian wilderness. It is not at all surprising to find these conflicting interpretations — de-extinction as the resurrection of a mammoth or the protection of the elephant — within the same texts. The title of Shapiro's book is far from the only implication it gives of having discovered a method for "successful de-extinction": "While others will undoubtedly have different thresholds for declaring de-extinction a success than I do, I argue that this — the birth of an animal that is capable, thanks to resurrected mammoth DNA, of living where a mammoth once lived and acting, within that environment, like a mammoth would have acted — is a *successful de-extinction*, even if the genome of this animal is decidedly more elephant-like than mammoth-like" (Shapiro 14; emphasis added). Yet, there are just as many intimations that justify the project as elephant conservation: "Our goal is to turn an elephant into an animal that can survive in cold places" (Shapiro 46). This kettle logic is capable of harnessing various energies an advocate for "de-extinction" might hope to attach to their work; on the one hand, pragmatic and familiar goals no serious conservationist could eschew, on the other hand, exciting, novel, science-fiction-become-reality innovation.

Omnimpotence

"De-extinction" has come to the fore at a sort of crossroads for conservation work. As Ben Minteer describes in his elegant study of pivotal questions facing the field today, *The Fall of the Wild*, the realities of climate change have displaced the foundations of the field. If the very name *conservation* suggests the task of preserving what has been, an existing ecological balance, it is no longer the case that one can count on a future that will resemble the past — it seems increasingly necessary to intervene in ecosystems, to translocate species, reengineer, or hybridize them, in order for them to keep pace with their changing environs. In a certain sense, the emergence of "de-extinction" at this moment of crisis or double bind is no accident — at least to the extent that it is an emanation of the very technoscientific cultures whose way of life has brought about this instability. That is undoubtedly a broad categorization, that would include all of our contemporary ills as well as everything that resists them, as if from the inside. It is not to cast doubt on the field that I situate it in this context or political ecology. As I have tried to demonstrate, it is difficult and likely impossible to define *de-extinction* without incorporating much that is old together with what may be new, and none of these techniques or technologies can simply be discarded or disregarded in the face of our responsibilities today.

Nonetheless, it is within this framing that a certain complicity comes to light. De-extinction is often described as a techno-fix or techno-optimism in response to the imminent threats of climate change. This description is apt, not because we can or should discard "de-extinction" in favor of something non-technological, as if such a thing existed, but because it represents a certain attitude toward or within that totality or rather contextuality that I just hastily summoned or sketched.[9] As I have tried to demonstrate, the question of species is a *practical* question. This is precisely an effect of its arbitrariness — if there is no truth or necessity that can guide us, then it is always necessary to ask why and how, toward what end, a particular definition or intuition of "species" or anything else has come to guide the living in this and all endeavors. If the power to create and re-create the "species" were under our command, it would be as if there were no more extinction. The finality and irreversibility that has belonged to extinction since it was first recognized as a possibility would be lifted up. Relieved and relived, there would be life at will. And yet. These projects that proceed fitfully over the course of years, how much control could they possibly gain over extinction if it claims several species a day at present? And if many extinctions are a result of the disappearance of habitats and ecosystems that would be necessary to resettle a "de-extinct" species? From this perspective, many of the pragmatic, conservation-minded justifications for the elaborate "de-extinction" projects seem no better than rationalizations. I say this without any intention of denying the potential value of these technologies and projects and even the ethical imperatives that may demand them — it is only that the context within which they seem to represent a power over life is overdetermined by more powerful contexts that I am trying to awaken here. At the very least, a greater caution is in order with respect to how the project is discussed, to obviate the implications of a genetic determinism that brings both the knowledge of life and power over it entirely within our command.

Similar to the optimism surrounding carbon-capture technologies, there is a risk that even the prospect of "de-extinction" would appease our sense of responsibility while continuing the very practices that have created the urgent threats we pretend to absolve ourselves of. To question our knowledge and control of life (or species) is not a defeatism or discouragement from acting, but a call to confront a more vast responsibility that is all the more pressing for being inexhaustible. Something is lost at every turn, or remains over without belonging to another site or another knowledge, conscious or unconscious, of a life and species that has always been what we do not know. Or rather, like the *gone* introduced in the first chapter, it is no more lost than gained, the only promise and possibility of a future resides in this non-present remanence. It calls for what is no more conservation than transformation. There is no way of preparing a site for it — for what never takes place in or as a place — or preserving what dispenses with itself, without risking at the same time snuffing it out. And there is no such thing as inaction — certainly not within the world we have created, which is not the one we want precisely because everything is now a matter of our will (our omnimpotence). Not that we can control it of course, but that nothing will any longer transpire that is not codetermined by our commission or omission. Yet, this is only one figure of the impossibility of inaction, an impossibility that runs much deeper, that will outlast even this "civilization" and its resistances, for the gone is just as much what holds and releases *us*, what we do not know of ourselves, what receives us when we come and dismisses us when we go. Our "own" extinction would be anything but a satisfaction of this responsibility, which demands that we seek our "selves" no less than the other, in the place of other, the gone.

A deconstruction of our ecological responsibility is not merely to heed the call of the question concerning technology, nor to recognize that where the danger is, there grows the saving power, but to face a necessary dis-juncture within all power, possibility, essence, being, and responsibility (Derrida, *Specters* 1–61). This dis-juncture is our relation of non-relation to the inaccessible secret from which we receive every definition or de-limitation of species — which compels us to decide both what it has been, is, and will be, while leaving us with inexhaustible undecidability.[10] If we are enticed and compelled by the lure and allure of species, this or that and all species, if losing one seems like a terrible guilt and bringing one back like absolution even though we have never known what that is, and have only arbitrarily set its limits, that can only be because we receive it as an inheritance from the other. This différance or trace, as referral to the other, that makes even our most self-assured science possible and impossible, places a limit within everything. It is this limiting de-limitation that I fear certain advocates of "de-extinction" most hope to obscure. I sense this aspiration toward infinity, toward infinite life, everywhere that a spiritual and ontotheological vocabulary animates their self-aggrandizement. They routinely refer to their task not just as "de-extinction" but as *resurrection*: one de-extinction project has named itself the Lazarus project, and another the Ark project.[11] Moreover, its practitioners often recognize the appeal of promising to expunge a guilt incurred by the human species by focusing their efforts on species such as the passenger pigeon, which were lost due to unimaginable acts of our exterminating violence (passenger pigeons once numbered in the billions).[12] Stewart Brand, who was the editor of the *Whole Earth Catalog* and has been one of "de-extinction's" staunchest advocates since cofounding Revive & Restore, expresses its messianic promise explicitly: "Humans killed off a lot of species over the last 10,000 years. Some resurrection is in order. A bit of redemption might come with it" (Brand). This "redemption" is always and only possible where the very finitude that has driven what we call "life" from before the beginning is abolished, where an infinite or spiritual life can be volatilized or rarefied from it.[13] Where one can pay off a debt such as extinction in kind, in specie. Now more than ever it is necessary to tarry with the inadequacy of this exchange — it is not too late to discover something there that is not simply a loss.

Cauda

> Phage tails are molecular machines that specifically recognize bacterial host cells, penetrate the cell envelope and deliver the phage genome into the cytoplasm. All tailed phages [. . .] belong to the order of *Caudovirales* (from "cauda," which is Latin for tail).
>
> — Nobrega et al.

It is nothing like an error or passing whimsy that led biologists to craft Trees of Life they knew to be impossible, nor to situate viruses on the vertical or horizontal passages of those trees, that led virologists to search for an origin and patriarch of their field, that led Venter to invoke an essentiality underlying his research and life in general, nor that led those involved in de-extinction research to affirm or deny a certain essence of species. There is no simple fault here because one cannot not speak or act as if something like essence guided one's practices. The lure of necessary and impossible essence makes every honest scientist seem at once to speak in cunning or bad faith, or to be the dupe of a time, place, or paradigm they could not see for what it was. One must always take certain precautions before rushing into any such judgments, because there is no demystified or disillusioned position outside this structure, no more here than anywhere else. If I have tried to rethink "life" from its undecidability with certain operations attributed to virality, that is certainly not to suggest a simple exit from the essentiality of reproduction or the reproduction of essentiality. There is every risk and even a certain necessity of "the virus" becoming an essence in turn, and of its being unthinkable or unspeakable unless some essence of life precedes it. The task of deconstruction is not to step outside of this necessity, which is no more physical than metaphysical, but to think it without guilt or fault (not as a fall or lapse from pure being, nature, or self-identity). No one is to blame for what fails without fail, and no one is ever outside of the fray or diffraction of its ungrounded drift. Deconstruction would not be to dismiss everything that has been attributed to the accomplishments of science or technoscience, from some supposed neutral superiority, but rather to think this necessary impossibility as the condition of possibility of everything that has come to pass as science or life.

Parasite, Parricide: Family Scenes

There is no life without some virality, and thus no pure life, no spontaneous or autonomous selfhood. In the same breath, it is necessary to say that there has never been a true virus. "Virus" has never been a concept, has never belonged to a theory or paradigm, and no virus has ever been observed, contacted, or contracted. Far from disinfecting our corpus, from leaving us free to build our family trees or phylogenies or to live our lives, it is only on this condition that what we have called virality and thought we could recognize under that name has come to pass. Occupying any given form or concept only as a borrowed site, achieving its most cunning ruses precisely because there was never any thing, substance, or being there to receive the returns. This is undoubtedly what is most strange, most foreign to scientific or philosophical conceptuality, and perhaps — depending on one's tastes — most frightening or disgusting, unsettling or uncanny, about what I have been calling virality (a name that is as provisional as any other identifying mark), that it is not *opposed* to life, that it makes what we call the life in us and our "self"-reproductions possible, even as it makes any pure selfhood or pure reproduction impossible.

Everywhere that something like a family or filiation takes form, a viral volatility threatens to displace its foundations. Those biologists who hope to place themselves within a vast phylogeny or genealogy, a Tree of Life, find their efforts destabilized by "horizontal" thefts or gifts for which the virus is often the model or vector. Those virologists who hope to identify the founder or father of their discipline find the instability of virality uproots even this notion of institution, origin, or discovery. At the same time, everything that has ever been counted or counted itself as a family has only been possible on the basis of something like virality. What depends on reproduction or iterability exposes the same or the claims of ownership or propriety, of true lineage as against intruders, to the risk and chance of what may not return, of what confronts the self or same with an alterity unrecognizable and unreclaimable as its own. Only where this otherness is possible can the re- of re-production or re-clamation take place. Thus, it is not because some concrete entity with a special power of subversion happens onto this scene that what we call virality threatens life "itself." There has been life and its "reproductions" only because that virality was there before the beginning, making its reproduction possible even as it made pure reproduction impossible. "The" virus, then, is a sort of comforting fiction for one who would like to believe that there was some pure life that could be preserved against a chance intruder. Virality overtakes the virus's given forms as well. It is necessarily the case that this term also remains provisional, that it can be overtaken by structures or setups in which its usage here is hardly recognizable.

A "viral fossil" (the EVE) that is as essential as our "proper" genes for human reproduction is only one of the eventualities that the virality of vitality makes necessarily possible. For those synthetic biologists who sought to harness the power of reproduction, creating a kind of technological graft onto a cyborg family tree, it often appeared that they nonetheless maintained quite traditional notions of their own parentage. They impressed upon themselves all the more their role and responsibility as father figures as they tried to torque life outside of its "natural" families (Venter proudly celebrating Synthia's birthday, or lab technicians protesting that they are owed a kind of filial duty, love or respect, from the clones they help create). It would not be so surprising to us to find this family form imposed on what seems most foreign to it, if we recognized that the family has been from the start a compromise and negotiation with structures of substitutability, virtuality, phantasm, alterity, and reading — in a word, virality. Nor should we be so surprised to find ontotheological drives that inscribe each of these scenes within figures from Genesis (the Tree of Life, Eden, Adam and Eve, and Noah's ark, for example) — every family is, from the first, a spiritual family.

Everything the biologist calls horizontality is related to what is sometimes called queer kinship, which can quite literally take the form of a spread by viral contagion. Tim Dean, in his *Unlimited Intimacy,* frames his study of the "subculture" of barebacking by explicitly contrasting it with the effort to normalize gay culture by assimilating it to vertical heteropatriarchal forms of filiation: "My thesis is that the emergence of a subculture of bareback sex is not merely coincident with but directly related to the campaign for same-sex marriage that has occupied so much attention in recent years. [. . .] [G]ay men have discovered that, on the basis of viral transmission, they can form relations and networks understood in terms of kinship — networks that represent an alternative to, even as they often resemble, normative heterosexual kinship" (ix–x). Note that Dean acknowledges both how horizontal kinship is a sort of by-product of efforts to conform to vertical structures of genealogy, and the remainder of some undecidability or *resemblance* between the two. Nothing prevents these horizontal relations from conforming at the level of desire or fantasy to traditional, vertical, familial structures, as Dean notes repeatedly throughout his study. For example:

> Viral transmission facilitates fantasies of connection, kinship, and generation that may be as familiar to heterosexual as to lesbian and gay readers. The fantasy of bearing someone's child or, indeed, of becoming someone's child is not gender specific or a function of sexual orientation. To those who might object that the joyful outcome of heterosexual insemination could not differ more from that of HIV transmission, we must reply that the unconscious remains blind to any such difference. Although the social outcome differs, the fantasy motivating both kinds of reproduction is essentially the same. (Dean 87)

Given that the most seemingly proper and "natural" structures of vertical filiation have always already been phantasmatically imposed or doubled, the risk or promise is ever-present that our horizontal networks will simply become another affiliative structure of inclusion and exclusion, which only means that it makes possible new forms of parasiticity. Frequently, in Dean's study, one finds the most seemingly disseminal sexual relations overlaid with fantasies of monogamy and genuine parentage.

We should note as well the mechanisms of filiation that function in Dean's text to account for this horizontal drift or *dérive*, which should not merely be attributed to "rhetoric," as if there were a physical or biological realm independent of it: "The AIDS epidemic has given gay men new opportunities for kinship, because sharing viruses has come to be understood as a mechanism of alliance, a way of forming *consanguinity* with strangers or friends. Through HIV, gay men have discovered that they can 'breed' without women" (6; emphasis added). A virus, a parasite, can make you a parent, can give you the power of childbirth, of a-filiation, the production of common blood — it is a supplement that makes one belong to a family of one's own (something that is never simply a "natural" given). Thus, even where Dean seems to oppose viral "consanguinity" to the family structure, he nonetheless finds these viral transmissions constantly inscribed in family trees. Dean speaks preferentially of becoming siblings in a "bug brotherhood" or becoming someone's "Daddy" by "having fathered his virus" (85). We can note the biblical overtones of this scene, which recalls Jesus's exhortations to despise what we sometimes call the natural family, precisely in order to accede to a spiritual family figured in the same terms. A Christian culture has shared in the emphasis on masculinist filiations, figuring God as a father and the spiritual community as a brotherhood or fraternity — just as the twelve disciples included only men.

Moreover, precisely because the family is *originally* a phantasmatic structure, virally imposed and exposed, it is not a straightforward matter to say that the mother is absent from these scenes (despite Dean's emphasis on fatherhood), nor even that breeding happens "without women." As Derrida wrote, "the mother (whatever forename or pronoun [*prénom ou pronom*] one gives it) is situated beyond sexual opposition. It especially isn't a woman. She allows herself simply to be represented, detached, by a sex" (Derrida, *Clang* 153; right column). Far from the claim that there is one woman and one man in every act of filiation, that is simply to say that this supposed division or opposition, sexual difference, forms a fold or pocket within each of us, and runs through or across all the participants in the most heterogeneous encounters.[1] What gender is the virus? Virality is never simply virility.

If we tried to compile the countless ways viral figures have snuck or been conscripted into family scenes, we would find something far more chimeric. One mark of the virality of vitality: the oldest meaning of *virus* in the English language (and one of the meanings of its Latin etymon) was *semen*. The first microscopists who were able to image viruses by means of the electron microscope found recourse to a similar theme when attempting to analogize their vision. With a surprising regularity, as if it were no accident, tailed bacteriophages (caudaviruses) were described as "sperm-shaped" or "spermatozoon-like" (Luria et al. 57; Kriss 621).[2] Giuseppe Penso was perhaps more circumspect, comparing the phage to a "syringe," and offering the following, vivid description of its "swollen" and "turgid" tail: "Once the phage has been adsorbed on to the host, its tail pierces the cell wall and penetrates either into the protoplasm or into the virtual cavity between the wall and the cytoplasm, but the penetration is in any case very small and involves only the swollen part of the tail. For a short time after adsorption the phages retain their normal appearance, but quite soon a loss of thickness and turgidity becomes clearly observable and they also become more transparent to electrons. There is a definite impression of some passage of matter from the phage to the host" (Penso 254). Such observations even led Max Delbrück and Salvador E. Luria, both of whom received Nobel prizes for their work on virus genetics, to posit a still stronger analogy between the virus and sperm. In an article that frequently repeated the "sperm-shaped" description of tailed phages, they noted that the infection of a cell by a virus seemed to prevent other viruses from crossing its membrane, leading to what Delbrück called their "penetration hypothesis": "[I]n multiple infection, most of the adsorbed virus particles do not penetrate the cell. One might assume, therefore, that the entrance of the first virus makes the cell wall impermeable to other virus particles, *just as* the fertilization of an egg by one spermatozoon makes the egg membrane impermeable to other spermatozoa. This may be termed the 'penetration hypothesis'" (Delbrück 166; emphasis added). Beyond mere analogy, this observation led them to suggest something about the filiation of viruses: "An interpretation of this kind, [. . .] would suggest an analogy with the fecundation of monospermic eggs, and would lend support to those theories of the systematic position of virus which consider it as related to the host rather than as a parasite" (Luria et al. 66). This chain of associations, linking sexuality to autonomy or property (the virus as "related" to the host, the endogenous theory), and therefore to non-parasitism, is remarkable. Everything we have read confirms its fragility.

On the other hand, French scientists who were among these early microscopists offered imagery no less sexual in nature, but where the linkage between virility and virality was far less straightforward. For instance, "The head of this bacteriophage is slightly ovoid[;] it [*elle* — referring to *la tête*] possesses a tail that can be wavy and whose extremity seems swollen."[3] Another group of observers described a veritable primal scene: a "spherical head" and a "short tail" with a "swollen extremity," all "enveloped by a sheath" — that is, a *gaine*, from Latin *vāgīna* (Giuntini et al., "Étude" 790).

A series of stunning and quite recent discoveries, that have required the redrawing of the borders of virality, and therefore of life as well, have provoked recourse to a similar imagery. The viral family whose original distinguishing mark was only its smallness relative to the cell, has expanded to include giant viruses, sometimes called *giruses* (Van Etten et al.). These monstrosities contain significantly larger genomes than any previously known virus, with genes that are considerably different from those found in the rest of the living world (so much so that it complicates any theory that would derive viruses from cellular life) (Colson et al.). While some bacteria have only about 200 genes, viruses have been discovered with as many as 2500 (Philippe et al.; Arnold). The first girus to be discovered was named the mimivirus, short for *mi*micking *mi*crobe, because it resembled a Gram-positive bacterium under a Gram stain (La Scola et al., "Giant").[4]

Researchers have continued to discover and classify ever larger viruses, creating a curious family. After mimivirus, they dubbed a still larger girus mamavirus, presumably because it was, in the colloquial sense, a mother of a virus.[5] An ironic name, given that, if we follow those who exclude viruses from the Tree of Life, a replicator like a virus can have no mother. Still, the maternal nomenclature has persisted: following mamavirus, two genera of giruses have been christened *Pandoravirus* and *Pithovirus*.[6] The original Pandora myth gives a curious account of the origin of maternity. Elissa Marder, in *The Mother in the Age of Mechanical Reproduction* and "Pandora's Fireworks," reads the Pandora myth as an expression of anxiety over the technicity and alterity inherent in reproducibility. Pandora, who introduces humankind to maternity in the myth, is not herself natural, but a product of *technē*, of art and technology. At a time when, we are told by Hesiod, only men existed, born directly of the earth, Pandora is shaped by Hephaestus from water and clay to introduce into humanity women and childbirth. (This progression from autochthony to sexual reproduction is preserved in the creation story of Genesis as well, where Adam, whose name suggests earth [*adamah*], is created from the "dust of the ground," and where Eve [*Hawwa*], whose name suggests life [*haya*] and who is called "the mother of all living," condemns humanity to childbirth [Gen. 2:7, 3:20].) The jar known in English as Pandora's Box is in Greek a *pithos* (the etymon of the monstrous *Pithoviruses*), an earthenware jar with curved handles that can be read as a figure for the womb. Pandora, who is herself a technical product, disseminates her gifts, her seeds or germs, by means of a technical prosthesis of the womb. The myth places prosthesis at the origin — the birth of birth itself is not natural, but technical, cultural, manufactured. Like a virus that uses the other (life) as its machine.

Shortly after the emergence of SARS-CoV-2, an infectious disease physician and two virologists wrote a short call to action entitled "Escaping Pandora's Box." Their misreading or misremembering of the Pandora myth can only be a symptomatic resistance to the threats of artificiality that it depicts as anything but accidental faults: "The Greek myth of Pandora's box (actually a *pithos*, or jar) comes to mind: the gods had given Pandora a locked jar she was never to open. Driven by human weaknesses, she nevertheless opened it, releasing the world's misfortunes and plagues" (Morens et al.). This notion of a trusting God betrayed by human frailty is far more Christian, imposing the story of the Tree of Knowledge onto the Greek myth. There is perhaps human frailty in Hesiod in that Epimetheus was instructed to be wary of the tricks of the gods, and to refuse gifts from Zeus, but Pandora, the woman-machine, seems to have been made precisely to spread ill. That being said, both stories depict life as originally without sexual difference (or at least without sexual knowledge) and without suffering or death.

I propose to add a member to this family of maternal replicants — knowing full well the risks of creating a term or germ that may not only develop a life of its own, but a family and species of previously unimaginable monstrosity. Of course, there is no non-creation, and my silence or reticence could just as easily feed the ghostly spread of an epidemic. I do not pretend to be able to control or study its movements within the confines of what one could in good faith call an experiment, though perhaps just as much can be learned from it if it renders the limits of experimentability legible. I call it *mimevirus*, and it lives inside you now. It makes no difference whether we say that it has no body, or that it is our body, the bridge and master that we now live in service of (to the extent that we or anything lives in or through us). Our relationship may well prove symbiotic or commensalist; it may ask nothing of us, even dwell silent and forgotten within us, and yet enable the otherwise impossible. In a way, the mutualist symbiont is simply the most effective parasite. It is as easy as whispering its name, or even trying not to think about it, to spread it to another, or to oneself as the other. As hysterical identification demonstrates, even the unconscious is contagious. Yet, having no given body (even its name admits of translation, paraphrase, and every effect of substitution and dissemination), the network of its contagion and any attempt at eradication is a matter of faith. Nature, as imitation or mimesis, shares its origin and its possibility with this mimevirus — it is not clear that one could "cure" it without extinguishing life itself, and perhaps it signals the very impossibility of an ultimate end to life (an immortality one could just as easily lament as celebrate). Far from something that arrived late on the scene, it is only a cultural or decadent creation if we understand cultivation and decline, degeneration (*Entartung*), to have constituted the possibility and impossibility of the origin — and to haunt the interiority of anything that might come to be recognizable as a corpus.

Notes

Prodrome

1. See Flint et al. 3; Raoult and Forterre 315. Though the truncated form of this phrase does not occur there, Salvador Luria's 1950 article on the bacteriophage is sometimes pointed to as an early form of the "obligate intracellular parasite" definition (507). Lwoff's "The Concept of Virus" is also considered canonical, where the phrases "strict parasite" and "strictly intracellular" appear, and where a set of criteria is given for the virus in general: "(1) possessing only one type of nucleic acid, (2) multiplying in the form of their genetic material, (3) unable to grow and to undergo binary fission, (4) devoid of a Lipmann system" (243, 246). Aspects of this definition have been challenged by recent discoveries (Raoult and Forterre; Colson et al.). The idea that the virus only reproduces inside its host's cells already appeared in Beijerinck's 1898 study, to which I return in chapter 3. For an expansion upon the idea of virus as symbiont, see Grasis.
2. Koonin's statement about the incomplete translation system still stands, though in the time since he published this definition in 2011, more genes related to translation have been found among newly discovered giant viruses (Rodrigues et al.).
3. I say "much closer to life itself" because Kant is emphatic that he is not attempting to describe *life* in the "Critique of Teleological Judgment," but purposive arrangement that can be found in nature itself. That is, for Kant "life" implies something like a spiritual force or principle operating upon nature (upon a body or matter), whereas he is concerned with the organization and thus purposiveness that can be seen (but whose causality remains undecidable) in nature itself. Thus, his definition applies not to what he would call "life" but to organized beings, though it is strikingly similar to many modern attempts at defining life. See Basile, "Kant's Parasite."
4. See Tsing; Haraway, *Staying*; Latour, *Facing*; Malabou; cf. Basile, "Epic of Genesis."
5. I thank Thomas Clément Mercier for his help in translating this passage.
6. I have elsewhere written about the tendency, among authors associated with the "nonhuman turn," to misread poststructuralist authors as "humanists" by refusing to recognize that they approached language as something other than human. See Basile, "Misreading Generalised Writing"; Basile, "Political Climate."
7. I am not arguing that there is anything like a silence in Derrida's corpus on life, science, or any other theme relevant to my concerns here. For the most obvious example, see the recently published seminar *Life Death*.

Introduction

1. In short, their goal was to be able to say that they had created "authentic" poliovirus without relying on a cell (Wimmer S4).

2. The title of their initial paper used a similar phrase, but did not claim it as the first such accomplishment: "Generation of Infectious Virus in the Absence of a Natural Template." An earlier experiment used a similar process to reconstruct a virus but received relatively little attention (Blight et al.; Church and Regis 74). The first *in vitro* generation of a virus from a "natural template" took place in 1967, when researchers used DNA polymerase to copy the DNA of a bacteriophage, Phi X 174, and succeeded in creating or recreating infectious virus (Goulian et al.).

   Three scientists coauthored the paper describing results of the initial experiment in 2002 (Cello et al.). I will frequently attribute certain interpretations of those results to one of those three, Eckard Wimmer, primarily because the most theoretically far-reaching interpretations come from a later reflection he authored alone. He makes clear, however, that his original coauthors wanted to include much of this interpretive reflection in the original paper but were prevented by the editors of *Science* (Wimmer S5). Additionally, many of the main points are re-elaborated in another paper Wimmer coauthored with one of his two collaborators Aniko V. Paul.

3. Wimmer perhaps does not explicitly take this step himself, but it is implicit in many of his formulations, especially that of the ineradicability of viruses. Compare, for instance, "we did not 'create' poliovirus but merely reproduced a virus following the blueprint of the viral genome that existed in nature for thousands of years" (Wimmer and Paul 590). And, "Viruses have been defined as obligatory intracellular parasites in need of a living cell for replication. Our work challenges this definition. After the sequence of the poliovirus genome was determined, and a protocol for the *de novo* test-tube replication of the virus in a cellular extract was developed, the chemical synthesis of the genome was a logical step to assert its character as a chemical. For many virologists, the dual nature of *viruses as chemicals with formulae stored in data banks* and as organisms circulating in nature was not news. To most scientists and lay people, however, the reality that viruses can be synthesized was surprising, if not shocking. Then and now, we consider it imperative to inform society of this new reality, which bears far-reaching consequences" (Wimmer S9; emphasis added). Subsequent virologists have been more full-throated in articulating a definition of viruses that identifies them with their "coding capacity" (Raoult and Forterre 316).

4. Recently, nematodes that were thirty to forty thousand years old were discovered in Siberian permafrost and resumed their vital functions after being defrosted (Shatilovich et al.). A thirty-thousand-year-old giant virus was similarly "resurrected" (as the headline in *Nature* put it) — which raises curious questions about whether what has never lived (according to many biologists) can be brought back to life (Yong, "Giant Virus"). On the work by synthetic biologists to construct the genome of a living cell, see chapter 4.

5. Edward C. Holmes explains the commonplace confusions surrounding the term *quasi-species* in the study of RNA viruses (87–103). What Wimmer has referred to as the "average" genetic sequence of poliovirus Holmes would refer to as a "consensus sequence," which records the most frequently occurring nucleotide at each site in the viral genome (averaging out the mutation or mutations that are likely to occur in every copy). The quasi-species theory, however, was formulated to describe not the mere fact of a high mutation rate in a given population, but a specific type of population that would undergo natural selection differently because of its high mutability: the central figure of the quasi-species would not be the consensus sequence, but what Holmes calls the "'master' sequence," which may not be the most frequently occurring in the population, but is the fittest because its fitness is augmented by the tendency of other genotypes to mutate into it (Eigen and Schuster; Domingo et al.). While it is undoubtedly the case that RNA viruses have relatively high mutation rates, it remains an open question for virologists whether any viruses mutate at a sufficient rate to undergo such quasi-species dynamics.
6. The title of one of Wimmer's reflections on this experiment referred to the synthesis of "a Chemical *Called* Poliovirus" (emphasis added).
7. Definitions of disease eradication and extinction vary, but one widely used classification defines eradication as the global cessation of transmission among humans, and extinction as the complete elimination of the pathogen, from "nature" and from the laboratory (Dowdle). Thus, smallpox has been eradicated, but is not extinct, because stockpiles remain in high-security laboratories in the US and Russia. The best compromise the World Health Organization could devise during the Cold War was to divide the remaining stockpiles of the virus between the two countries, who have refused to part with them despite calls to do so from the WHO and many scientists (Kritz; Meyer et al.). In the case of poliovirus, there are not only laboratory stockpiles of the virus, but also transgenic mice genetically engineered to have cellular receptors for the virus, who pose a risk not only of spreading the disease in laboratories, but of spreading their gene to wild mice, allowing the disease to circulate among wild populations (Dowdle and Birmingham).

   The campaign for poliovirus eradication has faltered largely due to distrust and instability fueled by the US military and intelligence agencies. Poliovirus remains endemic in Afghanistan and Pakistan. There are several causes for this, including most recently a slowdown of vaccinations due to COVID precautions, but the root cause remains the destabilization of the region by decades of US warfare (international health care workers have been accused of gathering information to target drone strikes, for instance) and particularly the use of a fake vaccination campaign by the CIA to identify the whereabouts of Osama Bin Laden ("How the CIA"; Gostin; Kennedy; Cousins; Martinez-Bravo and Stegmann).
8. The United States signed the Biological Weapons Convention, internationally banning biological weapons, in 1972. Even if one trusts that their "dual-use" research will never be weaponized, the security of the oppressor may be a threat to the oppressed, a reality that may be obscured if one describes this research as purely defensive.

9. I hope to disturb the common sense according to which the dissemination of a virus would be an act of "bioterrorism," but the immense violence of the status quo, which includes countless biohazards in its distribution of waste, trash, environmental degradation, and indeed viral exposure, is representative of a state we agree to call "peace."

10. Gain of function research includes, for example, serial passage of viruses through nonhuman cell lines to see if they mutate to infect human cells. Justifications for such research include the argument that discovering the number of mutations necessary for an existing virus to become a threat to humans alerts us in advance to future pandemic risks (because one can say, for example, that a population of X species in Y region carries a virus that is only Z mutations away from possibly infecting humans). While performing such research on "possible pandemic pathogens" seems particularly shortsighted (Lipsitch and Galvani), it is nonetheless the case that certain undecidabilities haunt the attempt to ban a category or type of research. Many virologists have pointed to marginal cases where something that looks like it could be called "gain of function" research nonetheless seems to pose little risk and offer outsized benefits (Brouillette).

11. Wimmer attempts to account for this state of affairs, though he faces an aporia at every step: "Ultimately, the outcome of an experiment in fundamental research in biology, perhaps in all sciences, cannot be predicted. Biological systems are too complex to be able to account for all variations, no matter how carefully the experiment is designed. Experiments with viruses are no exception. Intuition and creativity play as great a role in science as they do in the arts, but are no guarantee of success. Thus, even the best scientists suffer frequent humiliation when their experiments do not yield the expected results. However, unanticipated results are not solely disappointments but have also led to landmark discoveries that have transformed an entire discipline or spawned new technologies. Indeed, unanticipated results might have revolutionized the practice of medicine — consider the serendipitous discovery of penicillin, for example.

These considerations explain why it is difficult to separate research projects that potentially pose a danger to society from the overwhelming majority of studies from which we draw benefits, including added security. Ultimately, scientists themselves must take the responsibility for evaluating the risks inherent in their research." (S8)

12. Perhaps this word, *deconstructs*, which is put to unfamiliar use here, standing without a direct object, could be heard as ergative or labile. That is, as a verb that, when it is not used in a transitive fashion, takes on something like a passive or reflexive meaning. In invoking this *lability*, I hope to avoid the self-reflexive connotations that were given to deconstruction by certain theorists who claimed it was something a text did to itself (Culler). I agree with Geoffrey Bennington that this reflexive usage suggests a self-present auto-affection, depending on notions of activity and substantiality that deconstruction ought rather to destabilize ("Deconstruction" 20). Vigilance is required to avoid the trap of either transitive or reflexive understandings of the verb. Deconstruction is the deconstruction of active and passive. So, I offer this turn of phrase, "the virus deconstructs," which can be heard as an ergative verb, as a sort of action of undergoing: it deconstructs, it breaks, it scatters, it disseminates — or, and this is not opposed to the first possibility, as an action without end.

13. I am most concerned here with the framing gestures of "materialist" work, both in its self-conception and in the political forces that impinge on it. Elsewhere, I have pursued a close reading of certain texts from this field, and the conflicts internal to them, to explore both what in them escapes the simplest representations of their commitments and how even the most apparently opposed gestures are capable of being domesticated within a given frame (Basile, "Life/Force"; Basile, "Other Matters").

14. I have focused on Jane Bennett's work as an example in what follows because it represents a relatively univocal statement of a tendency that seems to operate in much of the "materialist" field — even among those theorists who emphasize something like the non-self-identity of matter, its crossing of borders (E. A. Wilson; Barad; Haraway, *Staying*). Many of the authors who focus on this "entanglement" also draw explicitly from Derrida in articulating their theories of matter or nature (Schrader, "Responding"; Kirby; Schrader, "Microbial Suicide"). It is difficult to identify constants within this field, which is loosely recognized as part of a "material turn," and is capable of uniting disparate gestures, for instance: (1) the denunciation of past theory that overemphasized cultural and linguistic forces *or* the denunciation of those who separated nature and culture, coupled with (2) a turn toward matter and nature *or* a turn toward natureculture, the material-semiotic, entanglement, and so on. These contradictory or opposed gestures nonetheless give a similar form to the works in which they appear, and even coexist in individual texts. Similarly, an emphasis follows from these gestures on either ontology *or* something hyphenated like "onto-epistemology," again often within the covers of a single book. While some authors have attempted to sort these competing trends into a taxonomy of the field (Gamble et al.), I am less inclined to do so because I feel that in most if not all individual texts, one finds multiple strata competing or converging. Elsewhere, I have argued that the very declaration of a turn or epoch produces certain of the common effects that can be observed throughout the field (Basile, "Misreading Generalised Writing"; Basile, "New Novelty").

15. Cf. Basile, "Other Matters."

16. One could say, starting from the premises that what is universally given is natural whereas the cultural is particular to a place and time: "nature," because it is a given name in a given language, is necessarily a cultural thing. This is true no matter how many times we translate it, and so "culture" is what we find to be universal, and thus most natural of all. The point of such a demonstration is not to reverse a hierarchy, nor to provide a demonstration whose force remains within the domain of oppositional logic. The apparent conclusions of such a demonstration: *nature is culture, culture is nature*, depend on the traditional definitions of each term: nature as universal, culture as variant (tied to a specific people, place, history, etc.). If one would rather deconstruct this logic, then it is not a matter of demonstrating a reversal or contradiction within logic, but of showing that logic does not suffice to stabilize an opposition, thereby indicating what necessarily exceeds logic (without appearing as another language in another place, one that could oppose itself to logical language). Logic, too, is the most natural and the most cultural thing, a willful imposition occurring universally, something we can't help but invent. If the demonstration remains within this logical reversal, this can only take place if every other binary forming the architecture of logical thought is preserved

intact (universal/particular, necessary/contingent, production/inscription). Such a reversal leads to the commonplace lure of a statement such as "culture produces nature by naturalizing itself" (on the deconstruction of such formulae, see Kamuf). Rather than shifting the matrix of production from nature to culture, it is necessary to think how an event that belongs to neither could nonetheless only take place as their re-inscription. We will ultimately see that if life remains recalcitrant to logical definition, it is not because life is something natural and logic something cultural, nor vice versa, but because we are attempting, with this and every name, to grasp what exceeds this and every logic without there being anything else, something that would be an origin of itself.

17. The deconstruction of the life sciences, or biodeconstruction, has gained momentum in the last few years, thanks in part to the posthumous publication of Derrida's *Life Death*. See Timár; Vitale, *Biodeconstruction*; Lynes; Fritsch et al.; Kirby et al.; McCance; Mercier, "Resisting"; Senatore; Waltham-Smith; Mastrogiovanni; Rosenthal.

18. The promise to "believe in science" was already a rallying cry during Obama's first campaign — for an incisive critique of this rhetoric, see Dorothy Roberts's *Fatal Invention* (292–93).

19. Of the many proposals by geologists for how to date the beginning of the Anthropocene, I'll give one example that highlights how divisions *within* humanity were constitutive of its foundations. One proposed starting point for the epoch, 1945, the beginning of the nuclear age, is very possibly related to a certain nostalgia on the part of the baby boomers choosing the date (Latour, *Facing* 218n84).

20. For Capitalocene, see Moore, *Capitalism*; Moore, *Anthropocene.* For Plantationocene, see Haraway, "Anthropocene"; Mitman. For the racial Capitalocene, see Vergès; Karera; Saldanha. Cf. Clark, *Value* 20.

21. Timothy Morton gives a characteristically hasty and clumsy dismissal of such debates:

> "Quibbling over terminology is a sad symptom of the extremes to which correlationism has been taken. Upwardly reducing things to effects of history or discourse or whatever has resulted in a fixation on labels, so that using *Anthropocene* means you haven't done the right kind of reducing. But what if you are not in the upward reduction business? Scientists would be perfectly happy to call the era Eustacia or Ramen, as long as we agreed it meant humans became a geological force on a planetary scale. Don't like the word *Anthropocene*? Fine. Don't like the idea that humans are a geophysical force? Not so fine. But the two are confused in critiques of 'the anthropos of the Anthropocene.'" (Morton 20)

The circularity of this (self-)denunciation makes the ineluctability of inscription apparent. Morton would like to denounce "quibbling over terminology." In order to do so he must — quibble over terminology. His argument ultimately takes the form: "come now, we can all agree that what matters are things, not words, so let's use the word I prefer." How could it be otherwise? In order to say, what matters is not words, but things, he must pretend that the particular language he prefers can be elevated above this mere "quibbling," to stand as the proper name of "force" as such.

22. Increasingly, the focus of Anthropocene discourse is on placing its apparent unity in question, such as Steve Mentz's *Break Up the Anthropocene* or the 2021 conference Pluralizing the Anthropocene.

23. Authors in the present disagree over the relevance of discourses of overpopulation (Haraway, *Staying*; Clark, "But the Real Problem Is"; Merchant). I would note only that the question is not simply one of the referential value of a name, but of what politics we can realistically expect to form in the present around such a call to action. It seems to me that the eugenicist, xenophobic, and anti-immigrant right is far better posed to exploit its potential energy.

24. I will emphasize, in what follows, how the *appropriability* of any discourse, theory, thesis, position-taking, and so on overdetermines any attempt to derive political commitments from it. In Bennett's case, as with many other new materialists, her "vital materialism" argues that recognizing matter as a *living* thing will lead of itself to the correct ethical and political commitments. It is worth noting, then, that *carbon vitalism* has been a common strategy for climate denialists and their attempts to maintain the fossil fuel economy (Pasek; Malm and The Zetkin Collective 20). By calling carbon the "gas of life," they argue that only positive consequences will follow from its release into the atmosphere. However sinister or laughable we find this willful deception or self-deception, it nonetheless reminds us that vitalism *and any other thesis or position* cannot prevent its being appropriated for the worst.

25. That Bennett's political imaginary is typically limited to choices available to the individual subject as a consumer and voter manifests throughout her text. For example, she imagines the political efficacy of the slow food movement will come from its appeal to a peculiarly limited bloc: "What is distinctive about slow food, and what might enable it to become a particularly powerful assemblage, is its appeal both to the 'granolas' and to the 'foodies'" (Bennett 50).

26. Despite everything in her work that advocates for a form of relationality, Jane Bennett explicitly frames her project as the unification of a representation of power through the suppression of competing forms of solidarity: "Not Flower Power, or Black Power, or Girl Power, but *Thing-Power*: the curious ability of inanimate things to animate, to act, to produce effects dramatic and subtle" (6).

27. That inclusion has been as powerful a tool of oppression as exclusion, particularly in the history of antiblackness, is one of the major themes of Zakiyyah Iman Jackson's *Becoming Human*.

28. This possibility has been considered in work on ecological sadism and the death drive, though certain of its consequences are not drawn there (Malm and the Zetkin Collective 391–98, 502–06; Kornbluh).

29. It may be that psychoanalysis is one of few discourses where the consequences of this primary masochism are drawn. Nonetheless, there is always the risk that by naming this possibility, speaking of it for example as a or *the* death drive, will become an attempt to arraign it in the present rather than recognizing its haunting disjuncture at the heart of every experience and auto-affection. One finds both tendencies within Freud's work,

and thus every possible permutation among those who inherit from him, knowingly or unknowingly. Derrida's "Psychoanalysis Searches the States of Its Soul" poses the question of what sort of politics can follow from such a psychoanalysis.

30. There have been studies indicating a negative correlation between education and trust in the "scientific community" (Gauchat 727; Mirowski 175). Such surveys may suggest something like the operations of denegation. At the same time, they should not simply be taken as definitive. There are, as always, any number of ways to place in question such measurements, and nonfinite possible reframings of questions and studies that could always produce different results. There is never a positive science of the unconscious.

31. Despite vaccines being the only public health intervention Biden has continuously supported — to the point of dismantling interventions put in place by the previous administration — his commitment to vaccine distribution has privileged profits over public health, and has ultimately placed vaccine access to the hands of the market. The Biden administration has not made any realistic effort toward challenging the vaccine patents of drug companies or ensuring universal distribution beyond US borders (Adler-Bolton and Vierkant, "Pfizer"; Dearden). For the failures of the Trump administration, see Davis 26–34.

32. See Adler-Bolton et al.; Vierkant and Adler-Bolton, "Year I"; Vierkant and Adler-Bolton, "Year II"; Adler-Bolton and Vierkant, *Health Communism.*

33. It is important to note certain differences of context that account for how Derrida's discussion of "AIDS" differs from what one might say today. He refers throughout this interview to "AIDS" or the "AIDS virus" [*le Sida, le virus du Sida*], not making a distinction between HIV and the advanced stage of the illness it causes — perhaps this is more understandable in 1989, when no effective treatment existed to stop HIV from progressing to this stage. It would also be viewed as negligent, today, to refer to HIV as a "deadly contagion," given that safe and effective treatments exist for its management (though they are certainly not universally accessible), and that fearmongering about the disease is still used as a way of stigmatizing and criminalizing people living with HIV, particularly gay men (Thrasher). It is important to note, though, that when Derrida speaks of the "deadly contagion" depriving us "henceforth" of what the "relation to the other [. . .] could invent to protect the integrity" of the subject, he is not imagining anything like an untreatability of the illness, nor is he simply talking about disease in the organic sense. It is because a virus, an absent other that is neither living nor dead, is capable of transforming our relationship to the apparently living other and to ourselves, that we can no longer pretend that our self-relation is an integral whole — it is formed in advance by something like alterity or virality. This virality is just as symbolic as corporal, it makes the body or matter descend from what it is not. This remains true even once this particular virus becomes manageable, as it is today at least for those with access to medicine.

34. A universal that is at once a member of the class it is supposed to define, it has the status of what Derrida at times calls a "quasi-transcendental" (*Limited Inc* 127). What is perhaps the first occurrence of the term in Derrida's writing appears in *Clang*, though with a interestingly different gloss (183; left column).

Chapter 1

1. There have been several recent reflections by phylogenists on the need to include viruses within Trees of Life (Brüssow; Harris and Hill). Microbiologist Patrick Forterre uses a curious phrase when explaining his exclusion of viruses: "Viruses are not indicated in these trees but are intrinsically present because they infect the tree from its roots to its leaves" ("Universal Tree"; cf. Forterre et al. 557). Such a phrase admits of at least two readings, as does every virus of rhetoric and literality. "Infecting the tree" may refer to the *modeled*, the organisms represented by the Tree of Life whose history has been co-determined by what we call viral infections. It might just as well refer to the *modeling* model, the Tree of Life, which is diseased and denatured by something like a viral infection. We will see, later in this chapter, that the Tree of Life can neither be said to be nature itself, nor an artificial imposition on nature, and thus these referential and rhetorical readings could never settle into two separate and opposed possibilities.
2. One could perhaps make the image work if the tree grew from the center of the earth, with the tips of only those branches representing extant lineages reaching the surface, but Darwin's description of branches "decayed and dropped off" suggests that he is picturing the tree standing above the earth.
3. Despite certain appearances, the demand I am describing here, which is no more simply logical than natural, is not transcended by critiques of genetic reductionism. While the extreme forms of such thinking certainly correspond to a desire for a *simple* origin, the more networked and relational or plastic understandings of life, its development and evolution, are still reductions, despite everything else they change. Even if the possibilities of an organism and its lineage are not understood to follow directly from the presence of a given gene or genes, it is still the case, for there to be science or knowledge at all, that what changes and what remains the same must follow from a set of possibilities, of a plastic potential and its environmental channels. When a contemporary biologist speaks of the web of life, the hologenome, of developmental systems theory, plasticity, or niche construction, it is the case that the limits of a prior regime of causality are being transformed — but this is only possible by the imposition of new limits. The possibility must be understood to reside somewhere, however untraditional its locus, and that place must be manipulable to meet the demands of hypothesis and experimentation without which there is no knowledge, but only the unanticipatable. Thus, the "origin" is a doubling or haunting of what is or appears to be, its iterability and virtuality, regardless of how simple or complex its representation.
4. It is noteworthy that the vertical/horizontal binary has been made use of in multiple subdisciplines of the life sciences and medicine, often with different meanings. The usage I focus on here, in which "vertical" refers to inheritance by reproduction (in the case of prokaryotes, by binary fission) is customary for microbiology. Curiously, however, microbiologists borrowed this pair of terms from parasitologists, for whom "vertical" refers rather to infections that are passed along through the germ line or from parents to children by other means (through blood, eggs, milk, etc.). "Vertical" in this case refers to a process of infection, not reproduction, though the opposition is faltering, as we will see in the next chapter.

5. For attempts to update or surpass the Tree of Life, in order to take account of horizontality in the history of life, see Halary et al.; Dagan and Martin; Ragan and Beiko; Doolittle.

6. Even authors whose project is to challenge the universal applicability of a Tree of Life model carve out exceptions for our own branches: "To be sure, much of evolution has been tree-like and is captured in hierarchical classifications. Although plant speciation is often effected by reticulation and radical primary and secondary symbioses lie at the base of the eukaryotes and several groups within them, it would be perverse to claim that Darwin's TOL hypothesis has been falsified for animals (the taxon to which he primarily addressed himself) or that it is not an appropriate model for many taxa at many levels of analysis. Birds are not bees, and animals are not plants. But in other taxa or at other levels, reticulation may be the relevant historical process, and nets or webs the appropriate way to represent what is a real but more complex fact of nature" (Doolittle and Bapteste 2048). Cf. Bapteste et al.

7. I hope to follow the present work with a deconstruction of contemporary evolutionary theory, focused on what is sometimes called the Extended Evolutionary Synthesis. For introductions to each of these relevant movements, see Jablonka and Lamb; Margulis and Sagan; Oyama; West-Eberhard; Gilbert and Epel; Lewontin.

8. There are signs in Plato's text that this definition may be meant facetiously or ironically, complicating the intentions of the rational project (Miller 29–33; Naas 38–39).

9. Though the concepts of homology and analogy are often attributed to Owen, he was only redeploying them as well. For a more comprehensive history see Hall "Introduction"; Panchen "Richard Owen and the Concept of Homology."

10. For a brilliant discussion of the misreading of Owen in the construction of this historiography, see Amundson (76–106).

11. This all too brief history, which is not meant to be exhaustive (perhaps it is rather typological), could be further complicated many times over. One example of particular relevance to contemporary evolutionary theory, and to which I hope to return elsewhere, is what becomes of the concept of homology in light of plasticity (West-Eberhard 485–97). Similarly, epigenetics has been argued to complicate the task of correlating genetic homology with morphology (Haig 210–11).

12. This, for instance, is Daniel Dennett's explanation or explaining away of the reality of qualia (369–411).

13. In response to an article on the history and future of network models of life, Patrick Forterre wrote a curious report, every word of which merits reflection. For instance, when his comment that "mammals are clearly a branch in the tree of animals, and not a peculiar form of retrovirus despite the fact that retroviruses and derived element[s] comprise up to 80% of their genomes" takes for granted the dependence of this conclusion on the *name* (Forterre, "Reviewer's Report"). The invocation of a *lumen rationale* or clear and distinct idea ("clearly") disguises the order of operations, which cannot possibly derive tree-belongingness from categorical identification, but always the reverse; we recognize our place in a tree because the categories such as "virus," "animal," and "mammal," have been posited, together with so many others. The sense of family

resemblance orienting these categories preconditions all analogy and homology. For just this reason, the undecidability I am describing between or within vertical and horizontal cannot possibly be yoked within what Forterre refers to as a "dialectical" synthesis.

14. Like any attempt to model horizontality, Raoult's "rhizome of life" presupposes vertically defined relationships (the very ones that are corrupted in principle by horizontality). The definition of his project makes clear that understanding life as a rhizome depends on knowledge of the heterogeneity of its *origins*: "In accordance with the theory on the evolution of human societies proposed by Deleuze and Guattari, I believe that the evolution of species looks much more like a rhizome (or a mycelium) [. . .] Emerging species grow from the rhizome *with gene repertoires of various origins* that will allow, under favourable environmental conditions, the multiplication and perpetuation of this species" (105; emphasis added). (Note the erroneous limitation of Deleuze and Guattari's theory to "human societies.") The rhizomatic figure Raoult sketches in this article is only capable of representing this diversity of origins by color-coding the *roots* of an organism's genes (their "various origins") "according to current classification" — that is, precisely the vertical genealogy the rhizome aspires to free itself of.

15. Nonetheless, they explicitly acknowledge the undecidability of tree and rhizome: "There exist tree or root structures in rhizomes; conversely, a tree branch or root division may begin to burgeon into a rhizome" (Deleuze and Guattari 15). The question is not whether a given author or authors are capable of stating this, but how it is put to work or put in play in a text or reading.

16. Panchen notes that the discrepancy between pre-evolutionary taxonomies and reconstructed phylogenies are a result of "taxonomic inertia," and that evolutionary explanations of preexisting categories functioned as "*post hoc* rationalizations" (*Classification* 125). I would only add that this "inertia" is an enabling constraint — every taxonomy and every science works with or through some inheritance, without which there could be nothing at all.

17. Certain biologists have criticized this practice of capitalizing the "Tree of Life," precisely because they hope to suppress its biblical resonances (Penny; Forterre, "Darwin's Goldmine" 4).

18. A brilliant study of the biological Tree of Life from the perspective of microbiology has been written by a historian of science who happens to have the name Jan *Sapp*. Sapp mentions another grounds for comparison between the Biblical story of Genesis and contemporary biology — Oparin's theory of the origin of life, in which the first cells roamed without competition through oceans full of nutritious molecules, with no concern but to be fruitful and multiply, has been called an "aquatic Garden of Eden" (Hawkins 328; Horowitz 16; Sapp, *New Foundations* 222).

19. One symptom of framing this aporia as an opposition between nature and logic is Doolittle and Bapteste's claim that tree thinking is "imposed" on a naturally reticulate nature, and thus that network models can capture the true form of nature (Doolittle and Bapteste; Cf. Doolittle). Rather, there is no nature that is not the vestige or trace of what appears as a relatively virtual or thematizable theorization — one can displace this structure, but never escape it.

20. For the same reason, it is never a simple matter to escape the dominion of this tree, even if it dethrones itself. Robin Wall Kimmerer gives one example of the complex interweaving threads of tradition and theory that must be gathered and rewoven by one who hopes to alter the heading or bearings of this tradition. It can never be as simple as *opposing*, for instance, "Indigenous wisdom" to "scientific knowledge" — such an opposition would only be possible on the basis of an arbitrating logical system that was capable of recognizing and maintaining the separation of these terms on the basis of a common root. It would be, precisely, another tree. The problem is not simply that one must make use of a colonizing common idiom in order to accomplish this dialogue, but that there has never been, on either "side," a single or unitary language or tradition. Thus, there are only the placements and dis-placements of a *translation or traduction* that takes the risk of inheriting otherwise. Kimmerer's *Braiding Sweetgrass*, which weaves together what she calls "Indigenous wisdom" and "scientific knowledge" (Kimmerer is a botanist and a member of the Citizen Potawatomi Nation), demonstrates that this transposition is not a simple move beyond the mythology and schemas of treelike inheritance, but a manner of reconfiguring them. She recounts the story of Skywoman, as shared by the original peoples of the Great Lakes, a sort of creation story that also features a Tree of Life (Kimmerer 3–10). Arborescent kinship is not eradicated by this story, yet it is always possible that it may be inherited otherwise — in Kimmerer's retelling, Skywoman falls, but the earth is experienced not as a punishment for a guilty misappropriation, but a gift of generosity. Humans are, according to this tradition, "the younger brothers of creation" (Kimmerer 9). Her daughters, drinking sap from a maple tree, are "suckled by Mother Earth" (Kimmerer 66). Such redistributions of kinship and its stories is never a guarantee that relationships will become more just or sustainable — it is precisely because a conceptual opposition cannot be instated between them that one can never simply transition from an acquisitive economy to the generosity of the gift, which is certainly not to suggest that what we call gifts should not be given or acknowledged.

21. It is necessary to think what this figure suggests outside of the idea of a fall or loss. There is a not-having, an otherwise-than-being, which is prior to any possession and cannot be thought of as a simple deprivation. Derrida contrasts a certain thought of the gift to the idea of "giving-what-it-does-not-have" as it occurs in Lacan and Heidegger in *Given Time: I. Counterfeit Money* (2n2, 159n28).

22. Derrida puts forward a similar formulation in *Specters of Marx*, which is sometimes abbreviated as "*To be* . . . means . . . *to inherit*" (67; Bennington, *Scatter 2* 1–3).

Chapter 2

1. "The deconstruction of logocentrism, of linguisticism, of economism (of the proper, of the at-home [*chez-soi*], *oikos*, of the same), etc., as well as the affirmation of the impossible are always put forward *in the name of the real*, of the irreducible reality of the real — not of the real as attribute of the objective, present, perceptible or intelligible *thing* (*res*), but of the real as the coming or event of the other, where the other resists all reappropriation, be it ana-onto-phenomenonological appropriation. The real is this non-negative

impossible, this impossible coming or invention of the event the thinking of which is not an ontophenomenology. It is a thinking of the event (singularity of the other, in its unanticipatible coming, *hic et nunc*) that resists reappropriation by an ontology or a phenomenology of presence as such. I am attempting to dissociate the concept of event and the value of presence. This is not easy, but I am trying to demonstrate this necessity, like the necessity of thinking the event-without-being. Nothing is more 'realist,' in this sense, than a deconstruction. It is (what-/who-)ever arrives [*(ce) qui arrive*]. And there is no fatality about the *fait accompli*: neither empiricism nor relativism. Is it empiricist or relativist to seriously take into account what arrives — differences of every order, beginning with the difference of contexts?" (Derrida, "As If" 367; original emphasis).

2. Derrida considers a related set of questions with regard to the relationship of deconstruction to truth and ethics in "Afterword: Toward an Ethic of Discussion": "For of course there is a 'right track' [*une 'bonne voie'*], a better way, and let it be said in passing how surprised I have often been, how amused or discouraged, depending on my humor, by the use or abuse of the following argument: Since the deconstructionist (which is to say, isn't it, the skeptic-relativist-nihilist!) is supposed not to believe in truth, stability, or the unity of meaning, in intention or 'meaning-to-say,' how can he demand of us that we read *him* with pertinence, precision, rigor? How can he demand that his own text be interpreted correctly? How can he accuse anyone else of having misunderstood, simplified, deformed it, etc.? In other words, how can he discuss, and discuss the reading of what he writes? The answer is simple enough: this definition of the deconstructionist is *false* (that's right: false, not true) and feeble; it supposes a bad (that's right: bad, not good) and feeble reading of numerous texts, first of all mine, which therefore must finally be read or reread. Then perhaps it will be understood that the value of truth (and all those values associated with it) is never contested or destroyed in my writings, but only reinscribed in more powerful, larger, more stratified contexts. And that within interpretive contexts (that is, within relations of force that are always differential — for example, socio-political-institutional — but even beyond these determinations) that are relatively stable, sometimes apparently almost unshakeable, it should be possible to invoke rules of competence, criteria of discussion and of consensus, good faith, lucidity, rigor, criticism, and pedagogy" (*Limited Inc* 146).

3. Jonah resists his vocation as prophet because he knows, if people heed his warning and mend their ways, God will have mercy on them and not bring the prophesied punishment, and Jonah will be considered a false doomsayer: "When God saw what they did, how they turned from their evil ways, God changed his mind about the calamity that he had said he would bring upon them; and he did not do it. But this was very displeasing to Jonah, and he became angry. He prayed to the Lord and said, 'O Lord! Is not this what I said while I was still in my own country? That is why I fled to Tarshish at the beginning; for I knew that you are a gracious God and merciful, slow to anger, and abounding in steadfast love, and ready to relent from punishing. And now, O Lord, please take my life from me, for it is better for me to die than to live.' And the Lord said, 'Is it right for you to be angry?'" (Jon. 3:10–4:4).

4. Increasingly today, virologists distinguish "the virus" from the *virion*, the icosahedral shell that still figures in many artists' renderings of a virus. When a textbook illustrates

"the virus," one typically sees either this shell, sometimes with a cutaway revealing its sinister contents, or a sort of flowchart displaying the viral "life cycle." Many authors have argued that the virus should be thought of more as a process, or its proper form associated with more diffuse stages of this cycle (Dupré and Guttinger). The virus when dispersed within a cell, or having formed its own viral factory there, can be thought of as the mature form of a living thing for which the virion is only a spore or seed (Bândea; Villarreal ix; Claverie). It has even been suggested that an infected cell becomes a virus (Forterre). These sorts of re-definitions or re-schematizations of a concept or figure can always have unanticipatable effects at the level of thought, scientific discourse, or even matter "itself." Nonetheless, whatever it is that prevents the "virus" or anything else from forming a simple vertical family of descent or even from straightforwardly being a horizontal contagion, what makes it most resemble what it is not will perhaps remain the most *viral* thing of all. In this, there is something sublime about the virus beyond simply its ultramicroscopic hiddenness — it is the unfigurable, what shatters every attempt of the imagination to picture it, no doubt producing a mixture of attraction and repulsion by its counterpurposiveness. Curiously, while Kant defined the sublime as the absolutely large, other authors have probed its limits in the direction of the microscopic or infinitesimal (Addison; Kant 5:248; Derrida, "Parergon" 136–37).

In suggesting this viral sublimity, I am going against one of Lwoff's convictions: "Some scientists visualize the virus as an ill-defined shape emerging bashfully out of a dense and golden cloud. This is a beautiful and romantic vision. Virology should, however, not be too Turnerian. Nor should it be an abstract art. The portrait of a virus should not produce an aesthetic emotion by means of an organic disturbance. The virus is amenable to intellectual analysis" ("Concept" 251).

5. I noted in the introduction that viruses at one point in their history were referred to as *principles*. Their transition from lysogenic to lytic cycle can be referred to with another term borrowed from the logician: *induction*.

6. What appears here as a circumstantial terminological infelicity shelters a deep undecidability of roles (as always). Debate over this terminology is, so to speak, a live issue, as one article from 2016 demonstrates: "The apparent equivalence of these two terms — terms that in 'Lytic or lysogenic' seemingly are used to imply polar opposites — stems from an error in usage. That is, rather than a description of a phage property, the term 'Lysogenic' when correctly employed is a description of a *bacterial* property. Specifically, certain bacterial cultures exist that when added to cultures of certain other bacterial strains can give rise to the lysis of these other cultures. The first culture thus has the property of being able to generate lysis in the second culture. This first culture therefore is lysis generating, that is, lysogenic" (Hobbs and Abedon 6). The uneasy or imperfect parallelism between lytic and lysogenic is attributed here to the fact that one term describes a power of the virus, and one describes a power of the bacteria — but it is never straightforward to say what power belongs to a parasite and what to its host.

7. The term *temperate* is a translation from the French *tempéré* (Jacob et al.). The mildness it suggests might perhaps be reconcilable with what Heidegger described as *die sanfte Zwiefalt der Geschlechter*, the soft or tender twofold of *Geschlecht*, of race, gender, and so on (171).

8. As soon as an alternative exists, one can search for causes that guide the "decision" or "fate" leading to one or the other. The alternative between lysis and lysogeny is no exception, and researchers have even found mechanisms that operate as a form of communication among viruses to determine their course (Erez et al.).

9. Superinfection does take place, but is considered rare (Fogg et al.).

10. On the relations of infection and heredity in the domain of immunity, whose history often centered around the question of whether cancer could have a viral origin, see Creager and Gaudillière.

11. He also noted the unsteady terrain of lysogenic immunity: "The fact that an organism may, along with its genome, perpetuate hereditarily a structure from which a virus can be produced is in itself remarkable. More remarkable still, this structure, this potential virus, is the very factor responsible for immunity towards the homologous true virus. Attention therefore is called to lysogenic bacteria which, together with lysogeny proper, perpetuate the solution — for them — and the problem — for us — of cellular immunity" (Lwoff, "Lysogeny" 328).

12. This term, *autocatalytic*, stems from what historian of science Robert Olby has called "the enzyme theory of life," which took the genetic substance to be protein. Muller was one of many researchers operating within this paradigm, and who sought to shed light on the chemistry of life by exploring the gene-virus analogy. Olby gives a wealth of examples (149–65). Cf. Lily Kay's "W. M. Stanley's Crystallization of the Tobacco Mosaic Virus, 1930–1940."

13. In 1929, geneticist J. B. S. Haldane wrote a speculative essay about the origin of life that took inspiration from Muller's analogy. Haldane wrote, "The main difference between such a lethal gene, of which many are known, and the bacteriophage, is that the one is only known inside the cell, the other outside . . . A simple organism must consist of parts A, B, C, D, and so on, each of which can multiply only in presence of all, or almost all, of the others. Among these parts are genes, and the bacteriophage is such a part which has got loose" (Haldane 6).

14. "Viro-tauto-logy" would have elicited Lwoff's disapproval. It is what linguists would call a *hybrid*, combining roots from Latin and Greek, and thus would be prone to a critique he levels against the term *provirus*: "*Previrus* would be more correct. *Pro* and *phage* are greek; prophage is correctly formed. *Virus* is latin and *provirus* is a greco-latin hybrid. But, as even some purists consider that latin and greek roots have become so acclimatized that they have become part of our languages, we will keep provirus, (a) because it is already in use; (b) because of its balanced phonic analogy with prophage" (Lwoff, "Lysogeny" 323). I dwell on these questions because this is the very movement of virality, by which the improper displaces the origin of the proper "itself."

15. When the fertility factor takes the form of a plasmid, a copy of it is transferred entirely to the recipient cell, who thus becomes a potential donor, while the original donor cell also retains a copy. This plasmid can also integrate itself into the bacterial genophore, in which case if conjugation lasts long enough, all of the bacterium's genes — except the fertility factor — may be transferred to the recipient. Such a bacterial donor is referred

to as an Hfr or high-frequency-recombination cell. The recipient remains unable to be a donor after conjugation with an Hfr cell. Still, the F factor typically spreads quickly through a population, while also facilitating the spread of other genes.

16. The "present" is never a single age. Everything ages at a different rate — whether one examines geology, life, language, or culture, changes do not take place at the same rate, things that have retained an ancient form stand alongside what has changed quite recently. Moreover, things can change while remaining the same, keeping something like their form while taking on an unprecedented function in a structure that has altered around them or within them. There is no simple means to calculate the relative age of things because there has never been a simple present. This is even or especially true of genetics: "Intragenic recombination has been observed in numerous genes, and gene-conversion events tend to make copies of duplicated genes more similar to one another [. . .]. The segments involved in intragenic recombination usually are less than a few hundred nucleotides in length [. . .] much less than the length of typical genes. As a result, different regions within a single gene may have different evolutionary histories" (Gogarten et al. 2231).

17. Villarreal argues not just that sex plasmids are "derived from phages" but that they persevere in their virality; they are "a component of the continuum of genetic parasites that we call viruses" (83).

18. Most estimates cite the 2001 analysis of the Human Genome Project, which concluded that 1.5% of the genome was protein-coding genes, and about 8% was virus-derived long-terminal repeat retrotransposons (Lander et al.).

19. Jonathan Wells gives a summary of these functions in chapter 6 of *The Myth of Junk DNA*. It is curious to compare his presentation of the subject to that of Nessa Carey in *Junk DNA*. Though they ostensibly have the same project, demystifying the idea that the majority of the human genome is useless "junk," Carey creates a clear division between useful junk, and harmful junk, among which she counts the EVEs (Carey 5, 37–44). This repression can only stem from an anxiety in the face of one's own virality. Cf. Bardini 29–96.

Chapter 3

1. Lwoff recalls this fact with an appropriate degree of amusement but is rather inventive in his historiography: "in A.D. 50, Cornelius Aulus [*sic*] Celsus produced this remarkable sentence: 'Rabies is caused by a virus'" ("Concept" 240). The original Latin reads, "Utique autem si rabiosus canis fuit, cucurbitula virus eius extrahendum est" (Celsus 112–13; V.28.2). Early significations of *virus* could refer to any slimy liquid (including, as I will return to in the "Cauda," semen) or to a venom or poison. As with most words, it developed by displacement and condensation, never restricting itself to "negative" connotations. For instance, drawing on *virus* as meaning a sharp taste, Pliny could advise that resin could "allay [the] virus" of a harsh wine, but that for a "flat, mild wine" one should "add a touch of virus" (qtd. in Hughes 110).

2. This could result in statements that appear, in our necessary hindsight, to wreak havoc with relations of genus and species: "We know nowadays that contagious or virulent affections are caused by small microscopic beings, which are called microbes. [. . .] The microbe of rabies has not been isolated as yet, but judging by analogy we must believe in its existence. To resume: every virus is a microbe" (Pasteur 289).

3. For clarity's sake, I have altered the translation of *Vermehrung*, which can indicate several forms of multiplication (the translator used *propagation* in this case, but elsewhere uses *reproduction*).

4. A reference in the OED from 1931 suggests that in English *virus* had attained its specialized meaning before anyone designated themselves as a virologist (1946), though these archives are never perfect indications of usage.

5. The title of his paper is "Concerning a *Contagium vivum fluidum* as the Cause of the Spot Disease of Tobacco Leaves."

6. In most instances where the translator employs the term *virus*, Beijerinck has used the cognate German word *Virus* — except in one instance earlier in this paragraph where it translates the German *Gift*, poison.

7. One need not limit the search for predecessors to just this example. Pasteur not only studied rabies in the decade prior to Beijerinck's "discovery," but was also able to develop a vaccine for it (Hughes 32). Nonetheless, he typically goes unmentioned in histories of virology (or is present only as part of the background explanation of germ theory), because he was hesitant to posit anything but a bacterial cause for the disease, perhaps because it could represent a challenge to his germ theory.
   Before Ivanowski, Adolf Mayer had also made important investigations into the causes of tobacco mosaic disease, and identified it as contagious in nature, but found, oddly, that its contagiousness could be eliminated by filtration, concluding that it was a "bacterial disease" (Mayer 24). His memorable description of his naysayers can serve as a warning to us here: "And finally, there are many who hold the disease to be entirely unexplainable [*ganz unerklärlich*], a sort of magic [*eine Art Hexerei*], and several times the warning cry [*Zuruf*] has reached my ear: You will never find it, never!" (Mayer 14n4).

8. Similarly: "Beijerinck was the first to give the name 'virus' to the agent responsible for tobacco mosaic" (Lecoq 931; my translation).

9. It would take until 2000 for Bos's own commemorative article to be published, "100 Years of Virology." There, he mentions in passing that attempts to mark the centennial in 1992 had been "renounced," giving a citation to his earlier essay (Cf. Horzinek, "Birth").

10. The appearance of the term *heterologous immunity* in d'Herelle's description may lead to some confusion. Today, that term is reserved for antibodies created in response to a specific antigen that nonetheless have some ability to fight another antigen. Thus, they are part of the body's "own" immune response, and not what d'Herelle is imagining here — immunity mediated by a foreign organism that renders the body's response unnecessary or supplementary (such usage of "*hétérologue*," to refer to a foreign body,

was typical at the time). Brock makes the mistake of reading this passage according to the term's current meaning (159).

11. For example, Sally Smith Hughes, writing *The Virus: A History of the Concept* in 1977, summarily dismisses the possibility of phage therapy, saying only "this proved not to be feasible" (85). Cf. Stent 8–9; Venter, *Life* 173–74.

12. D'Herelle later claimed that Bordet and Ciuca began their research by stealing a vial of infected bacteria from his laboratory while he was traveling. D'Herelle claimed that they were unaware the bacteria in this vial had acquired immunity to phage (were in our terms lysogenic), and that this was the reason for certain erroneous findings of their experiments. (Summers 194–195n12)

13. Summers (71–72) argues against this combative interpretation of Bordet's invocation of Twort, but it seems plausible to others, myself included (Duckworth 798).

14. At the risk of being somewhat elliptical, I will say that a certain symmetry between Latour and Husserl, both of whom promise a presuppositionless return "to the things themselves," perfectly illustrates Derrida's assertion that "abandoned to the simple content of its conclusions, the ultra-transcendental text will always look just like the pre-critical text." (*De la grammatologie* 61; Bennington, *Not Half* 63n11).
    This theme, that Latour's discourse is a simple extension or "translation" of the "actors" or "agents" it discusses, without anything like representation, works across the entire surface of his text, but is explicitly stated several times as well: "To speak of 'revolution' is difficult enough in politics, but it is impossible in such a subject. The temporal framework itself is useless. What makes the history of the sciences — so respectable elsewhere — usually disappointing is that it sets out from time in order to explain the agents and their movements, whereas the temporal framework merely registers after the event the victory of certain agents. If we really wanted to explain history, *we would have to accept the lesson that the actors themselves give us* [emphasis added]. Just as they made their societies, they also made their own history. The actors periodize with all their might. They give themselves periods, abolish them, and alter them, redistributing responsibilities, naming the 'reactionaries,' the 'moderns,' the 'avant-garde,' the 'forerunners,' *just like* a historian — no better, no worse . . . instead of explaining the movements [*le déplacement*] of the actors by time and dates, we would explain at last the construction of time itself on the basis of the agents' own translations" (*Pasteurization* 51).

15. Historian of science Simon Schaffer was also critical of what he called Latour's "hylozoism," Latour's readiness to speak for the will of the "nonhumans" themselves (Schaffer, "Eighteenth Brumaire"). Schaffer also notes many substantial differences between the English translation of *Les microbes* and the French original — the sum of these changes make it seem that Latour revised the text for the English translation (that they are not, or not all, carelessness or inventiveness of the translator). Perhaps Latour thought this revision was necessary in order to *Anglicize* the text, in the broadest sense.

16. How is this possible? That a theory responsible for so many scientific and medical successes could nonetheless be fundamentally in error? Or perhaps, even, otherwise than truth and falsehood, in a domain where that opposition is no longer pertinent, because

there is a certain fictionality that makes truth possible and impossible. According to germ theory, it is the infection of an organism by a given microbe that causes disease. Such a theory leads to the practice of germ warfare, based on the idea that microbes are enemies or invaders to be destroyed. Today, we are more aware that microbes are neither evil nor ill in themselves, but that *certain* microbes are, while others may be a potent defense against them, and even that an apparently self-same microbe may be harmful in one context, healthful in another. Thus, it is not *infection* itself, nor *germ* itself, that causes disease — even to the point that we may "infect" ourselves with good bacteria ("probiotics") in order to stave off the worst effects of eliminating other infections. Thus, the question of causality must shift to another level, why does this germ cause an infection while this other does not, why in this context and not that, or why is this infection bad while this other one is good, healthy, even producing effects of immunity?

17. Thomas Kuhn's later work offers a theory of scientific change as language change that could usefully be placed in dialogue with this theory of translation. It would not be possible here to elaborate the relationship between Kuhn and deconstruction or ANT, in part because, somewhat fittingly, there is no one Kuhnian theory but only a transition between paradigms in various stages of development. Posthumously published material such as "Commensurability, Comparability, Communicability" details his attempt to change from a theory of scientific "revolutions" to one of "language change." In short, neither of these theories is quite what I am describing here as the deconstruction of science or science as deconstruction. To indicate certain points that would be levers for deconstruction: first, it would be a matter not of locating "incommensurability" simply *between* scientific systems but *within* anything like a science or theory. Second, despite Kuhn's quite radical critique of the correspondence theory of truth, one still finds its basic structure as an anchor point in his thought — which perhaps grows more conservative in this later phase: "To learn any one of these three ways of doing mechanics, the interrelated terms in some local part of the web of language must be learned or relearned together and then *laid down on nature whole*" ("Commensurability" 44; emphasis added). Similar statements can be found in *The Structure of Scientific Revolutions*, though placed alongside claims that "revolutions" are not simply changes in language but in the "world" itself. It remains to ask what happens to a theory of science once it is recognized that "nature" and any other word we could assign to designate the outside of our system is nonetheless a structural element "within" it, and thus is just as prone to "revolution."

18. In a word, one grapples with the singularity of a text not by treating them all indifferently, but by tarrying with differences of contextuality to the point of exhaustion, where the impossibility of self-identity shines through.

19. Cf. "If the orator is heard as is, the network is decentered, even locally: there is no longer an intercepter, no longer a crossroads or intermediate; there is no longer a town; Hermes, the father of Pan, died on the Pentecost. A miracle, they say; such things don't happen. I can speak and hear from West to East; the walls come tumbling down from gusts of wind, from blares of music. I can have a relation directly to some object without an intercepter coming in between either to intercede or to forbid [*Je puis avoir rapport directement à quelque objet sans qu'un intercepteur s'interpose, j'ai relation ouvertement à l'autre sans qu'un intermédiaire s'intercale ni pour intercéder, ni pour*

interdire.]. Is the absence of a parasite so rare? Is immediacy so miraculous? Must the word always be a parable, that is to say, always aside, *para-* [*décalée*]? No. If it is not a miracle, can we build it?" (Serres 43).

20. Lwoff's "The Concept of Virus" is quite a playful essay, something that is lost when it gets cited as a canonical and standardized definition of the category. He concludes on something like the note I am suggesting here: "Belonging to an hyperlogical extrovert nation, I have coined numerous definitions as if I had really penetrated the essence of things. And I should not have discussed the intimate nature of viruses with more confidence if I had been myself a virus" (Lwoff, "Concept" 251).

21. Derrida comments on this dialogue at length in *The Politics of Friendship*, including a reflection on this (not simply) proper name (77–78).

Chapter 4

1. "On one side, in scientific publications and conference presentations (and especially grant applications), [artificial life] is compelled to justify itself in relation to the knowledge claims of theoretical biology, to which it is in danger of becoming a mere adjunct; on the other, its experiments in simulated evolution are often seen as merely useful new computational strategies in the field of machine learning or as new software and/or methods in the development of evolutionary programming. Inscribed in neither of these flanking discourses is the possibility of a potentially more powerful intrinsic narrative, to wit, that artificial life is actually producing a new kind of entity — at once *technical object* and simulated *collective subject*" (Johnston 13; original emphasis).

2. "Inspired by biological evolution, researchers in the field of digital evolution study evolutionary processes embodied in digital substrates. The general idea is that there exist abstract principles underlying biological evolution that are independent of the physical medium" (Lehman et al. 277).

3. Artemisinin provides a curious case study of both the power and false promise of synthetic biology. It is used as a treatment for malaria, a disease responsible for hundreds of thousands of deaths each year, mostly among children in the Global South. Until recently, supply was mostly derived from a plant, *Artemisia annua,* with high production costs and unstable yields. In 2004, the Bill and Melinda Gates Foundation offered fifty million USD in funding for a team promising to increase supply and lower costs of the drug by creating a synthetic biology pathway for its manufacture. Critics pointed out that this would harm traditional farmers of the plant, and risked further destabilizing the market if they cut back on growing because they anticipated being replaced by biochemists. As it happened, the worst of these fears did not play out, simply because the venture failed on one of its key promises. By the time it ramped up manufacture to industrial scale, selling its "semi-synthetic artemisinin" at cost was still significantly more expensive than deriving it from plants. Also, the pharmaceutical companies that purchased this chemical as a precursor saw the biochemical engineering company manufacturing it as a direct competitor with whom they did not want to do business. As is often the case,

a flashy new technology had attracted investments over more traditional possibilities that might have offered better returns; the fifty million dollars and estimated 150 person-years of labor poured into semi-synthetic artemisinin could have been expended on breeding higher-yielding strains of the plant or improving harvesting techniques. For a case study, see Dalziell and Rogers.

4. One direction of current research that could lead to a more genuine fulfillment of promises of sustainability through synthetic biology would be the industrial scaling of a pathway that would use the greenhouse gas CO_2 as its energy source (Scown and Keasling; Köpke).

5. This aspect is emphasized by several practitioners in the field, such as Andrew Ellington: "I'd say synthetic biology's key utility is to excite engineers, undergraduates and funding agencies. Its key disadvantage is to create hysteria in the defense community" and Jeremy Minshull: "Scientific progress is incremental, but people holding purse strings, public or private, are most excited by paradigm shifts and the prospect of quick payoffs. Synthetic biology, then, is a useful term to attract funding for the ongoing (~30-year-old) biological revolution, powered by advances in molecular biology techniques coupled with increases in computing power. It means whatever the listener wishes to hear" (Various).

6. There was just as much disagreement among Leduc's contemporaries as there is today over the true significance of his research. On its reception, see Keller, *Making Sense* 29–49.

7. Campos acknowledges Rob Carlson, who came up with this name, as one of the first to call synthetic biology *synthetic biology*. Carlson himself also acknowledges Steven Benner, who in 2003 published an article in *Nature* entitled "Synthetic Biology: Act Natural."

8. Cf. "We are survival machines — robot vehicles blindly programmed to preserve the selfish molecules known as genes" (Dawkins xxi).

9. "Software of life" descriptions appear in JCVI researchers' scientific papers as well (Hutchison et al., "Design" 1). There are certainly a long line of predecessors for this description of DNA-as-software (e.g., Moralee), though certain of Venter's elaborations upon the model push it toward its breaking points.

10. "During reductive evolution as a mycoplasma, many biosynthetic genes were lost and replaced by transporters residing in the membrane, resulting in a trade-off between these two categories" (Hutchison et al., "Design" 10). In a moment, I will raise some reservations about the generalizability of this practice of assigning individual functions to individual genes, but I will note here that researchers that have tried to explore the "essential genes of unknown function" in JCVI-syn3.0 have concluded that many of them may also provide transport functions (Danchin and Fang; Antczak et al.).

11. Several researchers with the Venter Institute have placed their work explicitly in the lineage of Morowitz's theories: "In 1984, physicist Harold Morowitz, recognizing that the mycoplasmas were the simplest cells capable of autonomous growth, proposed that these bacteria be used as models for understanding the basic principles of life (Morowitz

1984). Although the mycoplasmas are often called atypical bacteria because of their salient characteristics, for minimal cell purposes, the exact opposite is true. Mycoplasmas are excellent embodiments of what is constant in all bacteria and indeed in all cellular life" (Glass et al., "Minimal Cells" 3).

12. Cf. Roy 189–90. In *Molecular Feminisms,* Roy argues that work in "synthetic ecology," which studies the cooperative interactions of synthetic cells, advances beyond the "reductionism" of minimal genome research. Roy interprets this in Deleuzian terms, as a reductionism that "has gone so far into itself that it has nowhere else to go but back out, sending out new lines of flight" (186). I am trying here to contrast a Derridean thinking of undecidability with this Deleuzian style of thought, which ontologizes individuality in order to celebrate the movement beyond its hypostasized boundaries.

13. Cf. "If a synthetic genetic material can be designed to catalyze its own reproduction within an artificial membrane, life of a primitive type will have been created in the laboratory. Perhaps these cells will resemble the first forms of life on Earth, from nearly four billion years ago, but more likely they will represent something quite new" (Venter, *Life* 134).

14. In *Archive Fever,* Derrida explores the relationship of event and inscription when entertaining the question of how psychoanalysis might have developed otherwise if Freud and his interlocutors had access to the technologies of telecommunication we take for granted, especially e-mail: "I will limit myself to a mechanical remark: this archival earthquake would not have limited its effects to the *secondary recording*, to the printing and to the conservation of the history of psychoanalysis. It would have transformed this history from top to bottom and in the most initial inside of its production, in its very *events*. This is another way of saying that the archive, as printing, writing, prosthesis, or hypomnesic technique in general is not only the place for stocking and for conserving an archivable content of *the past* which would exist in any case, such as, without the archive, one still believes it was or will have been. No, the technical structure of the *archiving* archive also determines the structure of the *archivable* content even in its very coming into existence and in its relationship to the future. The archivization produces as much as it records the event" (*Archive* 16–17; original emphasis).

15. The unculturability of most bacteria has been called the "great plate count anomaly," and was recognized as early as 1932 (Staley and Konopka 324–25). It was first observed as a discrepancy between the plethora of microbes that could be observed by direct microscopic counting versus the relative paucity when plating (growing in culture) from the same aquatic samples. Today, it can be confirmed by what is sometimes called *metagenomics* — the gathering of genetic sequence information directly from environmental samples — that a similar gap exists between the species that we have cultured and the genetic diversity of prokaryotic life (Vartoukian et al.).

16. Philosopher of science Massimiliano Simons has written an excellent article that categorizes the conceptual difficulties facing minimal genome research and theory. I will note a difference with his analytic approach, which sees the role of the philosopher as the conceptual clarifier of scientific confusions: "Here, the philosopher can play a role by clarifying the conceptual confusions that are at work" (Simons 133). There is a risk to

picturing aporias as confusions admitting of disambiguation. There is no proper rational systematicity underlying these "confusions" such that they could be sorted into a stable logical architecture that would serve philosopher and scientist equally and without dissension.

17. Koonin still speaks about this combinatorial possibility as the substitution of individual genes that perform the same function.

18. The phenomenon of non-orthologous gene displacement, of genes that are unrelated or distantly related but correlate with similar functions in different cells, is prevalent among prokaryotic genomes (Koonin, "Comparative Genomics" 129).

19. "We refer to such a cell controlled by a genome assembled from chemically synthesized pieces of DNA as a 'synthetic cell,' even though the cytoplasm of the recipient cell is not synthetic" (Gibson et al., "Creation" 55).

20. It was found, about five years after the creation of JCVI-syn3.0, that its cells were abnormal with respect to morphology and cell division. The addition of seven more genes to its genome made it better resemble its natural prototype (Pelletier et al.). Several experiments have investigated the evolvability of this augmented "minimal" cell (Hossain et al.; Moger-Reischer et al.; Sandberg et al.), now dubbed JCVI-syn3A or JCVI-syn3B (Breuer et al.; Hossain et al.). Everything I am saying here is condensed in a passing remark in Moger-Reischer et al.'s conclusion: "From an engineering perspective, more studies are needed to evaluate the minimization of other genomes in alternate chassis under different environmental conditions. Nevertheless, if we assume that our findings are somewhat general . . . " (126).

21. Ultimately, Celera would also face the hard limit of the failure of its model of genetic agency. The promised medical applications of this information, which depended on the anticipation of discovering more single genes as causes of diseases and targets for therapies or screening, failed to materialize at anything like the promised scale. (Celera was ultimately sold to Quest Diagnostics for $8 a share in 2011.) Venter had already been let go long before this last curtain fell.

22. ExxonMobil no longer funds the project SGI/Viridos has been working on for over a decade to create fuel from chemical precursors generated by the photosynthesis of bioengineered algae. Chevron, United Airlines, and Bill Gates have since stepped in to support the continuation of the research (Elgin and Crowley, "Algae"). ExxonMobil spent tens of millions of dollars advertising this investment in green energy, which led several commentators to accuse it of greenwashing (Matthews; Elgin and Crowley, "Exxon"; Westervelt). Curiously, Exxon seems to have discontinued its funding as soon as Viridos started to show significant progress toward its objective.

As with every alternative fuel, algae biofuels come with tradeoffs. While the photosynthesis at the heart of this process would of course remove carbon from the atmosphere, energy would be required to maintain the algae farms and to process the lipids they produce into fuel. Ultimately, Viridos estimates that a commercial venture based on this technology would produce seventy percent fewer greenhouse gas emissions than conventional fuels. Additionally, it uses salt water to grow the algae, which places no strain on freshwater supplies, but its algae stocks still require significant amounts of fertilizer.

23. See Glass et al., "Minimal bacterial genome"; Venter et al.; Glass et al., "Installation."

24. Sociologist of science policy Jane Calvert has provided a brilliant reading of synthetic biology patents that explores the undecidability of their basic categories (172). While others write about jurisprudential concepts such as nature or abstract idea as if they were givens, Calvert recognizes the deconstructibility of these premises, and the resulting ethical or practical undecidability of their results. For example, while many denounce "genetic reductionism" as complicit in the biopolitical control over "life itself," Calvert observes that reductionist arguments have just as easily been put to use to deny such patents.

25. I am grateful to Avery Slater for making me aware of many of these connections.

26. While I appreciate ETC Group's wordsmithery, it is curious to note that they frequently embrace the same conceptual framework Venter uses to categorize his novelty — indulging in the same sort of positivism for the sake of fearmongering. ETC Group routinely refers to Synthia as "the world's 1st synthetic life form," in order to advocate for public oversight of the research (ETC Staff, "Men and Money").

27. Despite often casting itself as the historically aware and critical or skeptical counterbalance to scientific hubris, certain sociological discourses about synthetic biology participate in this epochal apocalypticism. Nikolas Rose's writing on the subject offers an interesting example, because he has passed through two almost diametrically opposed framings of synthetic biology that nonetheless participate in a deeper unity. His 2007 *The Politics of Life Itself* names synthetic biology as one of many examples of the advent of a new "age," in which "life itself has become open to politics" (Rose 15). Curiously, despite certain overtures it makes to skepticism about the self-representations of scientists and their economic motives, this epochal framing represents a profound dogmatism concerning the nature of biological truths. Rose must accept, for example, that interventions at the level of the genome are interventions into "life itself" or "our very biological life itself" (*Politics* 40), which means that he accepts synthetic biology's implicit definition of life and the model of finally revealed or achieved mechanism that he sometimes claims to see as a menace. The narrative he tells is in many respects familiar — this "new age" is one in which what used to have a proper belonging and self-identity (such as, for example, human reproductivity) is now substitutable, fungible, technical, prosthetic, synthetic, exposed to surrogates, and so on. It is not to pretend that nothing ever changes that I place this narrative in question, but simply to recognize that the non-self-identity implicit in modeling or in the model-reality system has always been part of what we call *life*, that this brings with it a technicity or prostheticity within which any value of the natural or vital is already caught up.

Thus, when Rose seems to take an opposite approach to "synthetic biology" just six years later, in "The Human Sciences in a Biological Age," the transition is not as great as it might appear. Here, Rose places more emphasis on the fictiveness of synthetic biology's claims of engineering control over life: "However, a closer examination of synthetic biology shows how misleading is this fantasy of biological control, and its foundational premise of life as pure mechanism" ("Human" 21). Rose explicitly names Venter's research into the minimal genome as an example of this mechanism. We can see that this attribution is, at best, half true. Despite Venter's explicit claim to have revealed life to be a machine and the many hallmarks in his work of a "reductionist" focus on isolated components of the organism, Venter is just as dependent on a certain vitalism that is never purely opposed to such mechanism. He *must* attribute these machine powers to "life" in order for them to function as powers at all. Now, Rose's objective in this essay is to *oppose* to each other a bad, ideological, and ethico-politically suspect mechanism, and a vitalist, non-reductionist biology with which he would like sociology to have an "affirmative" relationship ("Human" 23–24). If no biology, nor anything else, even "life itself," is ever purely mechanistic *or vitalistic*, this compromises in principle any such position taking and even the meaning of "affirmation." The reductionist model is part of life, and thus is not simply "reductionist" but rather pro-re-ductionist, what I have elsewhere called *traduction*. The concept of a vital presence exceeding the model is a function of the model, of différance. And Rose's invocation of "self-organization" as an example of vitalism to be affirmed obscures the aspects of modeling or difference that remain within any such science.

28. This term was first used in reference to an earlier, still relatively notional minimal genome published by Venter and other researchers (Hutchison et al., "Global Transposon Mutagenesis"; Reich).

29. It is not entirely clear from the description he gives whether the balloon was mentioned facetiously.

30. This phrase has appeared in the work of industry watchdog ETC Group in reference to Synthia, as well as the phrase "Pandora's Bug" (ETC Staff, "Patenting Pandora's Bug"). They use the term *original syn* with what is clearly a condemnatory and even damning tone. While I have voiced my concerns throughout this chapter about the economic and ethical questions raised by these experiments, I want to make clear that what I refer to here as an *original syn* is nothing negative.

31. See Shreeve 374; Potter. When Venter used this term, he was referring to a project that combined his minimal genome research with his aspiration to bioengineer a bacterium that could produce a renewable fuel source. He still operated under the pretense that his minimal bacterium would be the ideal chassis for bioengineering, a paradigm of knowledge and control. These projects continued in parallel, but Viridos's work designing a synthetic biological fuel source has not built upon the JCVI-Syn series of cells. Rather, it has started its work from algae, using precisely the trial-and-error method that Venter claimed his minimal genome research would supersede.

Chapter 5

1. Keller recognized an almost universal silence on this point; though this process was typically spoken of as "nuclear transfer," including in the scientific literature, the techniques going by this name (including the one that produced Dolly) often involve transferring much more than a nucleus into the enucleated egg cell (Keller and Ahouse). This technique would be more aptly described as cell fusion, and not enough is known to say

how much structures beyond the nucleus may be contributing to its success. Keller saw this reductive nomenclature as an extension of an ideology that attributed all agency in development to the nucleus and within it the genes. There is a similar technique that transfers only or almost exclusively the nucleus, but Keller is correct that there is a widespread silence on this difference, which obscures evaluation of the implicit models of agency (Wakayama et al.; Fulka et al.; Gouveia et al.).

2. Strictly speaking, this step is no longer necessary. A new method of "cloning" has been developed that can begin from a somatic cell, merely editing its epigenetic markers in order to return or turn it into an induced pluripotent stem cell (Boland et al.; Fan et al.). While there have been some successful cases of such a cell being implanted and developing into a new, genetically identical organism, this process still has a similarly low success rate (comparable to that of SCNT), again a sign that the underlying mechanisms are poorly understood (Takahashi and Yamanaka; Stanton et al.).

3. This planned collaboration between the mammoth de-extinction project and Pleistocene Park has been thrown into question by the political situation in Russia (Lamey).

4. An even older sample has been recovered using a different sequencing technique (Kjær et al.; Kjær and Willerslev). Two-million-year-old DNA sequence fragments were recovered not from a particular frozen organism but from an (also frozen) environmental sample analyzed with metagenomic sequencing. These samples could be matched with existing data to create a picture of the ecosystem existing at that time.

5. Beth Shapiro elaborates these difficulties in *How to Clone a Mammoth* (51–72).

6. "But what if — as we now see — the 'biologist' were no longer simply a biologist; what if in his or her work as a so-called biologist he or she had to do history, linguistics, semantics, chemistry, physics, the science of institutions, even literature?" (Derrida, *Life Death* 214).

7. For a journalistic introduction to Novak and his work, as well as Stewart Brand, Beth Shapiro, and others in de-extinction's orbit, see Rich.

8. Other conservationists offer species-definitions guiding their work that are open to the inherently ethical or moral dimension of defining species, and the role of a narrative or fabulative backstory, one in which we are interwoven, in determining what we are trying to save (Meine 12; Minteer 111).

9. Minteer recognizes the double bind in which these interventionist demands place us — which he expresses when weighing the moral and spiritual risks of a much simpler intervention than de-extinction: "But in my mind a bigger worry hangs over discussions about the risks and rewards of [assisted colonization] and related proposals that open the door to increased human manipulation and control of species and ecosystems. It's the objection that, although well intended, such efforts do not in the end address the deeper moral problem: the need to restrain ourselves on the landscape and, especially, to rein in our ecologically destructive activities. By putting us in a more commanding position in the natural world — which at the extreme end of the continuum promotes us to the role of 'planetary manager' — strategies such as AC can appear to elide this deeper moral challenge of environmental forbearance and

possibly even exacerbate an already dysfunctional human-nature relationship" (78). If our hope is to preserve some part of the world that would be beyond our control (what we call *nature*), we lose it in the very act of subjecting its conservation to our ecological and economic calculations and technological interventions. The displacement of thinking I am suggesting does not absolve us of this risk, but, setting out from its ineluctability, faces the necessity that the ethics of a "human-nature relationship" could only be justified by what seizes us unawares, a decision of the other that overtakes both halves of this impossible relation.

10. It is for this reason that no straightforward alternative is possible between the possibilities that structure Ursula Heise's discussion of de-extinction — whether it aims to restore something old or create something new (204–15). The legacy of what has been is always something we do not know, a task and a call to decision for us.

11. The Lazarus project aims at the "de-extinction" of a species of gastric-brooding frog, which is itself a sort of figure of rebirth; these frogs swallow their tadpoles and allow them to develop in their bellies, to regurgitate them as frogs. Beth Shapiro refers to this species as "the Lazarus frog," a name that may lead to some confusion. It has been described that way at times in media reports on the Lazarus project, though a number of other frog species claimed this title before it. A *Lazarus species* is a common term for species thought to be extinct that later reappear — there have been several such "Lazarus frogs."

12. There is debate over whether mammoth extinction was anthropogenic. Mammoths went extinct during a period of warming that began twenty thousand years ago, which may have been a factor in their decline. However, they had weathered similar periods of warming in the past, so it is possible that overhunting by humans that by then shared their territory was a fatal factor (Shapiro 1–5).

13. In each chapter thus far, the strange crossroads of necessity and contingency have met at the locus of certain "proper" names: Eugene, Sapp, Villarreal, Lysis (Socrates's companion), the *Apolutrōseōs* (a name of God), and Venter. I will admit that I first concluded this chapter without any such character to add to this collection or menagerie. I was saved, perhaps redeemed, by Antoine Traisnel, whose brilliant talk I heard at the 2024 conference of the American Comparative Literature Association, and who noted many of these biblical resonances in the self-justifications of "de-extinction." He also noted what had escaped me, hiding like a purloined letter in plain sight: that one of the most prominent biologists working in this field is named George *Church*. The Frozen Ark project describes itself as an effort to preserve, conserve, or save the DNA of endangered species, specifically by freezing tissue samples from specimens before they are extinct. The original Ark promised a future for life by preserving the reproductive potential of species. This "Frozen Ark" cannot necessarily claim the same potential for reproducibility, for instance if no recipient egg cell or surrogate womb remains to nurture this seed, or no environment to receive it. As Traisnel brilliantly explained, it seems as if freezing life, putting it to death, while sublimating it into knowledge, was being envisioned as a higher form of eternal life. Becoming unchanging, ideal,

or informational, scientific knowledge and power would be revealing its complicity with ontotheological sublation.

Cauda

1. It is rare to see sexual difference mentioned by those philosophers or scientists who attempt an explicit definition of life, though its very possibility often complicates terms such as *autonomy* or *self*-reproduction on which such definitions tend to rely (Keller, *Secrets* 113–43). Before one assumes, with some haste, that sexual difference must not be essential to life because the simplest organisms (and many others besides) reproduce without it, asexually, it is necessary to consider whether anything reproduces *without difference* — and whether there is any difference that is not, at least spectrally or phantasmatically, *sexual*. Far from imposing sexual difference as an opposition and binary (which is anything but a biological concept, if that means a concept derived from the biological sciences) on life as a whole, this would disseminate the possibilities of a sexuality that is always more and less than one. When living things unite and divide, it is never possible to know what facet of their being compels them. This secret mark or trace, that gathers the dissemination of individuals into a family or species and drives those individuals to reunite in patterns that reinforce or transgress these apparent divisions, is never ultimately known by an investigating scientist or a living thing itself. Is it a color, a smell, a gene, a symbiotic bacterium, that drives an individual to choose one mate over another?

In Derrida's "Ants" and Elissa Marder's "Insex," the figure of the in-sect, of something internally divided, represents this trace structure that is constitutive of the species and sexuality of the living. I object, however, to a condition and limitation Derrida places on the spectral expanse of this term: "There can be traces without sexual difference, for example for unsexed [*asexué*] living things, but there can be no sexual difference without traces, and this goes not only for 'us,' for the living thing we call human" ("Ants," 20–21). One can take on faith that perhaps, somewhere, there is something non-sexual that can still be considered a thing — but precisely because sexual difference is, in its most advanced or primitive form, something read from traces, one can never attest to a positive knowledge of sexuality's absence (there are no straightforwardly unsexed living things). Similarly, it troubles the good faith of our belief that there are living things without sexuality of any kind when we observe that even our own sexuality is not a positive knowledge of a thing that exists, but is our exposure to a différance we cannot master. For this reason, sexual education, research, and maturity or coming of age are lifelong pursuits. It is perhaps the case, in this instance, that Derrida was attempting to avoid the appearance of naïveté that might result from challenging scientific discourse and its distribution of sexuality and asexuality. However, his attempt to be circumspect introduces a more fundamental error. In *Life Death*, on the other hand, he is willing to challenge a biologist's discourse when it claims that there could be life without sexuality, that bacteria are asexual (Derrida 91–95; Vitale, *Biodeconstruction* 91–101; Vitale, "Microphysics").

2. These terms are put forward as English translations of earlier German and Russian studies (Ruska; Kriss and Tikhonenko).

3. "La tête de ce bactériophage est légèrement ovoïde, elle possède une queue qui peut être ondulée et dont l'extrémité semble renflée" (Giuntini et al., "Images" 580).

4. Technically, *Mimivirus* is a genus name, though its first discovered member, *Acanthamoeba polyphaga* mimivirus (APMV), is often colloquially referred to simply as "mimivirus" (La Scola et al., "Virophage"). Mamavirus also belongs to the genus *Mimivirus.*

5. This understanding of the name was confirmed for me by personal correspondence from Christelle Desnues, who also said the name was chosen for its resemblance to that of mimivirus. She cautioned that "we didn't imply any mother/daughter relationship with this name." It was originally explained as "we denoted this new strain mamavirus because it seemed to be even larger than mimivirus when observed by transmission electron microscopy" (La Scola et al., "Virophage" 100).

6. The discoverers of the *Pandoraviruses* explained their nomenclature thus: "These viruses are the first members of the proposed 'Pandoravirus' genus, a term reflecting their lack of similarity with previously described microorganisms and the surprises expected from their future study" (Philippe et al. 281). In a report on the discovery in *Nature*, the same sentiment was expressed with less reservation: "But these viruses, described today in *Science*, are more than mere record-breakers — they also hint at unknown parts of the tree of life. Just 7% of their genes match those in existing databases. 'What the hell is going on with the other genes?' asks Claverie. 'This opens a Pandora's box. What kinds of discoveries are going to come from studying the contents?'" (Yong, "Giant Viruses").

As for the *Pithovirus*, its discoverers made the connection with Pandora explicit, "Here we describe a third type of giant virus named 'Pithovirus' (from the Greek word *pithos* designating the kind of large amphora handed over by the gods to the legendary Pandora)[.]" (Legendre et al. 4274). A report in *Nature* added some details about its naming: "Evolutionary biologists Jean-Michel Claverie and Chantal Abergel, the husband-and-wife team at Aix-Marseille University in France who led the work, named it *Pithovirus sibericum*, inspired by the Greek word 'pithos' for the large container used by the ancient Greeks to store wine and food. 'We're French, so we had to put wine in the story,' says Claverie" (Yong, "Giant Virus").

Bibliography

Ackermann, Hans-W. "The First Phage Electron Micrographs." *Bacteriophage*, vol. 1, no. 4, July 2011, pp. 225–27, https://doi.org/10.4161/bact.1.4.17280.

Addison, Joseph. "The Spectator." *The Spectator*, vol. 3, no. 420, July 1712. Seventeenth and Eighteenth Century Burney Newspapers Collection.

Adler-Bolton, Beatrice, et al. *COVID Year Two*. https://soundcloud.com/deathpanel/covid-year-two-unlocked.

Adler-Bolton, Beatrice, and Artie Vierkant. *Health Communism*. Verso Books, 2022.

———. "Pfizer Walk with Me." *The New Inquiry*, 13 Sept. 2021, https://thenewinquiry.com/pfizer-walk-with-me/.

Aguilar, Wendy, et al. "The Past, Present, and Future of Artificial Life." *Frontiers in Robotics and AI*, vol. 1, 2014, https://doi.org/10.3389/frobt.2014.00008.

Alaimo, Stacy. *Exposed: Environmental Politics and Pleasures in Posthuman Times*. University of Minnesota Press, 2016.

Amundson, Ron. *The Changing Role of the Embryo in Evolutionary Thought: Roots of Evo-Devo*. Cambridge, 2005.

"And Man Made Life." *The Economist*, 20 May 2010. *The Economist*, https://www.economist.com/leaders/2010/05/20/and-man-made-life.

Anderson, Alun. "Craig Venter." *Prospect*, 22 Apr. 2006, https://www.prospectmagazine.co.uk/essays/57253/craig-venter.

Anderson, S., et al. "Sequence and Organization of the Human Mitochondrial Genome." *Nature*, vol. 290, no. 5806, Apr. 1981, pp. 457–65, https://doi.org/10.1038/290457a0.

Antczak, Magdalena, et al. "Environmental Conditions Shape the Nature of a Minimal Bacterial Genome." *Nature Communications*, vol. 10, no. 1, July 2019, p. 3100, https://doi.org/10.1038/s41467-019-10837-2.

Aristotle. *History of Animals. The Complete Works of Aristotle*. Edited by Jonathan Barnes, translated by d'A. W. Thompson, vol. I, Princeton UP, 1984, pp. 774–994.

———. *Parts of Animals. Parts of Animals; Movement of Animals; Progression of Animals*. Edited by T. E. Page et al., translated by A. L. Peck, Harvard UP, 1937, pp. 52–435.

Arnold, Carrie. "Were Giant Viruses the First Life on Earth?" *Quanta Magazine*, 10 July 2014, https://www.quantamagazine.org/were-giant-viruses-the-first-life-on-earth-20140710/.

Bândea, Claudiu I. "A New Theory on the Origin and the Nature of Viruses." *Journal of Theoretical Biology*, vol. 105, no. 4, Dec. 1983, pp. 591–602, https://doi.org/10.1016/0022-5193(83)90221-7.

Bapteste, Eric, et al. "Prokaryotic Evolution and the Tree of Life Are Two Different Things." *Biology Direct*, vol. 4, no. 1, Sept. 2009, p. 34, https://doi.org/10.1186/1745-6150-4-34.

Barad, Karen. *Meeting the Universe Halfway: Quantum Physics and the Entanglement of Matter and Meaning*. Duke UP, 2007.

Bardini, Thierry. *Junkware*. U of Minnesota P, 2011.

Basile, Jonathan. "The Epic of Genesis: Catherine Malabou and the *gêne* of Epigenetics." *Derrida Today*, vol. 16, no. 2, Nov. 2023, pp. 99–113, https://doi.org/10.3366/drt.2023.0311.

———. "Kant's Parasite: Sublime Biodeconstruction." *CR: The New Centennial Review*, vol. 19, no. 3, 2019, pp. 173–200, https://doi.org/10.14321/crnewcentrevi.19.3.0173. JSTOR.

———. "Life/Force: Novelty and New Materialism in Jane Bennett's Vibrant Matter." *SubStance*, vol. 48, no. 2, 2019, pp. 3–22, https://doi.org/10.1353/sub.2019.0018.

———. "Misreading Generalised Writing: From Foucault to Speculative Realism and New Materialism." *Oxford Literary Review*, vol. 40, no. 1, 2018, pp. 20–37, https://doi.org/10.3366/olr.2018.0236.

———. "The New Novelty: Corralation as Quarantine in Speculative Realism and New Materialism." *Derrida Today*, vol. 11, no. 2, 2018, pp. 211–29, https://doi.org/10.3366/drt.2018.0187.

———. "Other Matters: Karen Barad's Two Materialisms and the Science of Undecidability." *Angelaki*, vol. 25, no. 5, Sept. 2020, pp. 3–18, https://doi.org/10.1080/0969725X.2020.1807132.

———. "The Political Climate: Ecocriticism and the Apocalyptic Tone." *ELH*, forthcoming.

Bedau, Mark, et al. "Life after the Synthetic Cell." *Nature*, vol. 465, no. 7297, May 2010, pp. 422–24, https://doi.org/10.1038/465422a.

Beijerinck, M. W. "Concerning a Contagium Vivum Fluidum as Cause of the Spot Disease of Tobacco Leaves, in: Early Papers on Tobacco Mosaic and Infectious Variegation." *Phytopathological Classics*, vol. 7, 1942, pp. 33–52, https://doi.org/10.1094/9780890545225.001.

———. "Ueber ein Contagium vivum fluidum als Ursache der Fleckenkrankheit der Tabaksblätter." *Verhandelingen der Koninklijke Akademie van Wetenschappen te Amsterdam*, vol. VI, no. 5, 1898, pp. 3–24.

Bell, G. "The Sexual Nature of the Eukaryote Genome." *Journal of Heredity*, vol. 84, no. 5, Sept. 1993, pp. 351–59, https://doi.org/10.1093/oxfordjournals.jhered.a111356.

Benjamin, Ruha. "Black Skin, White Masks: Racism, Vulnerability & Refuting Black Pathology." *Department of African American Studies: Princeton University*, 15 Apr. 2020, https://aas.princeton.edu/news/black-skin-white-masks-racism-vulnerability-refuting-black-pathology.

Benjamin, Walter. "The Task of the Translator." *Selected Writings: Volume 1, 1913–1926*, edited by Marcus Bullock and Michael W. Jennings, translated by Harry Zohn, Belknap Press of Harvard UP, 1996, pp. 253–63.

Benner, Steven A. "Synthetic Biology: Act Natural." *Nature*, vol. 421, no. 6919, Jan. 2003, p. 118, https://doi.org/10.1038/421118a.

Bennett, Jane. *Vibrant Matter: A Political Ecology of Things*. Duke UP, 2010.

Bennington, Geoffrey. "Aberrations: De Man (and) the Machine." *Legislations: The Politics of Deconstruction*, Verso, 1994, pp. 137–51.

———. "Deconstruction and the Philosophers (the Very Idea)." *Legislations: The Politics of Deconstruction*, Verso, 1994, pp. 11–61.

———. *Not Half No End*. Edinburgh UP, 2010.

———. *Scatter 2: Politics in Deconstruction*. Fordham UP, 2021.

Birdsell, John A., and Christopher Wills. "The Evolutionary Origin and Maintenance of Sexual Recombination: A Review of Contemporary Models." *Evolutionary Biology*, edited by Ross J. Macintyre and Michael T. Clegg, Springer US, 2003, pp. 27–138, https://doi.org/10.1007/978-1-4757-5190-1_2.

Blight, Keril J., et al. "Efficient Initiation of HCV RNA Replication in Cell Culture." *Science*, vol. 290, no. 5498, Dec. 2000, p. 1972, https://doi.org/10.1126/science.290.5498.1972.

Bloor, David. *Knowledge and Social Imagery*. U of Chicago P, 1991.

Bojesen, Emile. *Forms of Education: Rethinking Educational Experience against and outside the Humanist Legacy*. Routledge, 2020.

Boland, Michael J., et al. "Adult Mice Generated from Induced Pluripotent Stem Cells." *Nature*, vol. 461, no. 7260, Sept. 2009, pp. 91–94, https://doi.org/10.1038/nature08310.

Bordet, Jules, and Mihai Ciuca. "Le bactériophage de d'Hérelle, sa production et son interprétation." *Comptes rendus des séances de la Société de biologie*, Masson et Cie, 1920, pp. 1296–98.

Bos, Lute. "The Embryonic Beginning of Virology: Unbiased Thinking and Dogmatic Stagnation." *Archives of Virology*, vol. 140, no. 3, Mar. 1995, pp. 613–19, https://doi.org/10.1007/BF01718437.

———. "100 Years of Virology: From Vitalism via Molecular Biology to Genetic Engineering." *Trends in Microbiology*, vol. 8, no. 2, Feb. 2000, pp. 82–87, https://doi.org/10.1016/S0966-842X(99)01678-9.

———. "One Hundred Years of Virology?" *ASM News*, vol. 61, 1995, pp. 53–54.

Bourgine, Paul, and Eric Bonabeau. "Artificial Life as Synthetic Biology." *Cyberworlds*, edited by Tosiyasu L. Kunii and Annie Luciani, Springer Japan, 1998, pp. 67–79, https://doi.org/10.1007/978-4-431-67941-7_5.

Brand, Stewart. "Opinion: The Case for Reviving Extinct Species." *Adventure*, 12 Mar. 2013, https://www.nationalgeographic.com/adventure/article/130311-de-extinction-reviving-extinct-species-opinion-animals-science.

Brannigan, Augustine. *The Social Basis of Scientific Discoveries*. Cambridge UP, 1981.

Breuer, Marian, et al. "Essential Metabolism for a Minimal Cell." *eLife*, edited by Zoran Nikoloski and Naama Barkai, vol. 8, Jan. 2019, https://doi.org/10.7554/eLife.36842.

Brinton, Charles C. "The Properties of Sex Pili, the Viral Nature of 'Conjugal' Genetic Transfer Systems, and Some Possible Approaches to the Control of Bacterial Drug Resistance." *CRC Critical Reviews in Microbiology*, vol. 1, no. 1, Jan. 1971, pp. 105–60, https://doi.org/10.3109/10408417109104479.

Brock, Thomas D. *Milestones in Microbiology*. Edited by Thomas D. Brock, Prentice-Hall, 1999.

Brouillette, Monique. "Gain-of-Function Research: Balancing Science and Security." *Hopkins Bloomberg Public Health Magazine*, 12 Apr. 2023, https://magazine.jhsph.edu/2023/gain-function-research-balancing-science-and-security.

Bruford, Elspeth A., et al. "Guidelines for Human Gene Nomenclature." *Nature Genetics*, vol. 52, no. 8, Aug. 2020, pp. 754–58, https://doi.org/10.1038/s41588-020-0669-3.

Brüssow, Harald. "The Not so Universal Tree of Life or the Place of Viruses in the Living World." *Philosophical Transactions of the Royal Society B: Biological Sciences*, vol. 364, no. 1527, Aug. 2009, pp. 2263–74, https://doi.org/10.1098/rstb.2009.0036.

Buecherl, Lukas, and Chris J. Myers. "Engineering Genetic Circuits: Advancements in Genetic Design Automation Tools and Standards for Synthetic Biology." *Current Opinion in Microbiology*, vol. 68, Aug. 2022, https://doi.org/10.1016/j.mib.2022.102155.

Callaway, Ewen. "Dolly at 20: The inside Story on the World's Most Famous Sheep." *Nature*, vol. 534, no. 7609, June 2016, pp. 604–08, https://doi.org/10.1038/534604a.

———. "Million-Year-Old Mammoth Genomes Shatter Record for Oldest Ancient DNA." *Nature*, vol. 590, no. 7847, Feb. 2021, pp. 537–38. *www.nature.com*, https://doi.org/10.1038/d41586-021-00436-x.

Calvert, Jane. "Ownership and Sharing in Synthetic Biology: A 'Diverse Ecology' of the Open and the Proprietary?" *BioSocieties*, vol. 7, no. 2, June 2012, pp. 169–87, https://doi.org/10.1057/biosoc.2012.3.

Campos, Luis. "That Was the Synthetic Biology That Was." *Synthetic Biology: The Technoscience and Its Societal Consequences*, edited by Markus Schmidt et al., Springer, 2009, pp. 5–22.

Carey, Nessa. *Junk DNA: A Journey through the Dark Matter of the Genome*. Icon Books, 2015.

Casali, Nicola. "*Escherichia Coli* Host Strains." E. Coli *Plasmid Vectors: Methods and Applications*, edited by Nicola Casali and Andrew Preston, Humana Press, 2003, pp. 27–48.

Cello, Jeronimo, et al. "Chemical Synthesis of Poliovirus cDNA: Generation of Infectious Virus in the Absence of Natural Template." *Science*, vol. 297, no. 5583, Aug. 2002, p. 1016, https://doi.org/10.1126/science.1072266.

Celsus. *On Medicine*. Translated by W. G. Spencer, vol. II: Books 5–6, Harvard UP, 1938.

Cepelewicz, Jordana. "The Hard Lessons of Modeling the Coronavirus Pandemic." *Quanta Magazine*, 28 Jan. 2021, https://www.quantamagazine.org/the-hard-lessons-of-modeling-the-coronavirus-pandemic-20210128/.

Church, G. M., and E. Regis. *Regenesis: How Synthetic Biology Will Reinvent Nature and Ourselves*. Basic Books, 2014.

Clark, Timothy. "'But the Real Problem Is . . . .': The Chameleonic Insidiousness of 'Overpopulation' in the Environmental Humanities." *Oxford Literary Review*, vol. 38, no. 1, July 2016, pp. 7–26, https://doi.org/10.3366/olr.2016.0177.

———. *The Value of Ecocriticism*. Cambridge UP, 2019.

Clark, Timothy, and Philippe Lynes. "Introduction: What Might Eco-Deconstruction Be?" *Oxford Literary Review*, vol. 45, no. 1, July 2023, pp. 1–20, https://doi.org/10.3366/olr.2023.0400.

Claverie, Jean-Michel. "Viruses Take Center Stage in Cellular Evolution." *Genome Biology*, vol. 7, no. 6, June 2006, p. 110, https://doi.org/10.1186/gb-2006-7-6-110.

Claverie, Jean-Michel, and Chantal Abergel. "Giant Viruses: The Difficult Breaking of Multiple Epistemological Barriers." *Studies in History and*

Philosophy of Science Part C: Studies in History and Philosophy of Biological and Biomedical Sciences, vol. 59, Oct. 2016, pp. 89–99, https://doi.org/10.1016/j.shpsc.2016.02.015.

Cohen, Haley. "How Champion-Pony Clones Have Transformed the Game of Polo." *Vanity Fair*, 21 July 2015, https://www.vanityfair.com/news/2015/07/polo-horse-cloning-adolfo-cambiaso.

Cohen, Jon. "Six Cloned Horses Help Rider Win Prestigious Polo Match." *Science*, 13 Dec. 2016, https://www.science.org/content/article/six-cloned-horses-help-rider-win-prestigious-polo-match.

Coleman, J. Robert, et al. "Virus Attenuation by Genome-Scale Changes in Codon Pair Bias." *Science*, vol. 320, no. 5884, June 2008, p. 1784, https://doi.org/10.1126/science.1155761.

Colson, Philippe, et al. "Reclassification of Giant Viruses Composing a Fourth Domain of Life in the New Order Megavirales." *Intervirology*, vol. 55, no. 5, 2012, pp. 321–32, https://doi.org/10.1159/000336562.

Cooper, Melinda. *Life as Surplus: Biotechnology & Capitalism in the Neoliberal Era*. U of Washington P, 2008.

Cousins, Sophie. "Polio in Afghanistan: A Changing Landscape." *The Lancet*, vol. 397, no. 10269, Jan. 2021, pp. 84–85, https://doi.org/10.1016/S0140-6736(21)00030-1.

Creager, Angela N. H., and Jean-Paul Gaudillière. "Experimental Arrangements and Technologies of Visualization: Cancer as a Viral Epidemic, 1930–1960." *Heredity and Infection: The History of Disease Transmission*, edited by Jean-Paul Gaudillière and Ilana Löwy, Routledge, 2001, pp. 203–41.

Csiszar, Alex. *The Scientific Journal: Authorship and the Politics of Knowledge in the Nineteenth Century*. U of Chicago P, 2018.

Culler, Jonathan D. *On Deconstruction: Theory and Criticism after Structuralism*. Cornell UP, 1982.

d'Herelle, M. F. *The Bacteriophage: Its Rôle in Immunity*. Translated by George H. Smith, Williams & Wilkins, 1922.

———. "On an Invisible Microbe Antagonistic to Dysentery Bacilli. Note by M. F. d'Herelle, Presented by M. Roux. Comptes Rendus Academie des Sciences 1917; 165:373–5." *Bacteriophage*, translated by Hans-W. Ackermann, vol. 1, no. 1, Jan. 2011, pp. 3–5, https://doi.org/10.4161/bact.1.1.14941.

———. "Sur un microbe invisible antagoniste des bacilles dysentériques." *Comptes rendus hebdomadaires des séances de l'Académie des sciences*, vol. 165, 1917, pp. 373–75.

Dagan, Tal, and William Martin. "Getting a Better Picture of Microbial Evolution En Route to a Network of Genomes." *Philosophical Transactions of the Royal*

Society B: Biological Sciences, vol. 364, no. 1527, Aug. 2009, pp. 2187–96, https://doi.org/10.1098/rstb.2009.0040.

Daggett, Cara. "Petro-Masculinity: Fossil Fuels and Authoritarian Desire." *Millennium*, vol. 47, no. 1, Sept. 2018, pp. 25–44, https://doi.org/10.1177/0305829818775817.

Dalziell, Jacqueline, and Wendy Rogers. "Are the Ethics of Synthetic Biology Fit for Purpose? A Case Study of Artemisinin [Point of View]." *Proceedings of the IEEE*, vol. 110, no. 5, May 2022, pp. 511–17, https://doi.org/10.1109/JPROC.2022.3157825.

Danchin, Antoine, and Gang Fang. "Unknown Unknowns: Essential Genes in Quest for Function." *Microbial Biotechnology*, vol. 9, no. 5, Sept. 2016, pp. 530–40, https://doi.org/10.1111/1751-7915.12384.

Dan-Cohen, Talia. *A Simpler Life: Synthetic Biological Experiments*. Cornell UP, 2021.

Darwin, Charles. *Notebook B*: [Transmutation of Species]. Edited by John van Wyhe, *Darwin Online*, 1837–1838, http://darwin-online.org.uk/content/frameset?itemID=CUL-DAR121.-&pageseq=38&viewtype=side.

———. "Online Variorum of Darwin's *Origin of Species*." *Darwin Online*, 2009, http://darwin-online.org.uk/Variorum/.

Davis, Mike. *The Monster Enters: COVID-19, Avian Flu and the Plagues of Capitalism*. OR Books, 2020.

Dawkins, Richard. *The Selfish Gene*. 30th anniversary ed., Oxford UP, 2006.

de Lisa, Ilaria. *The Patentability of Synthetic Biology Inventions: New Technology, Same Patentability Issues?* Springer, 2020.

Dean, Tim. *Unlimited Intimacy: Reflections on the Subculture of Barebacking*. U of Chicago P, 2009.

Dearden, Nick. *Pharmanomics: How Big Pharma Destroys Global Health*. Verso, 2023.

Delamater, Paul, et al. "Complexity of the Basic Reproduction Number (R_0)." *Emerging Infectious Disease Journal*, vol. 25, no. 1, 2019, p. 1, https://doi.org/10.3201/eid2501.171901.

Delbrück, M. "Interference between Bacterial Viruses: III. The Mutual Exclusion Effect and the Depressor Effect." *Journal of Bacteriology*, vol. 50, no. 2, Aug. 1945, pp. 151–70, https://doi.org/10.1128/jb.50.2.151-170.1945.

Deleuze, Gilles, and Félix Guattari. *A Thousand Plateaus: Capitalism and Schizophrenia*. Translated by Brian Massumi, U of Minnesota P, 1987.

Dennett, Daniel C. *Consciousness Explained*. Little, Brown, 2017.

Derrida, Jacques. "Ants." *Oxford Literary Review*, translated by Eric Prenowitz, vol. 24, no. 1, July 2002, pp. 19–42, https://doi.org/10.3366/olr.2002.003.

———. *Archive Fever: A Freudian Impression*. Translated by Eric Prenowitz, U of Chicago P, 1996.

———. "As If It Were Possible, 'Within Such Limits' . . . ." *Negotiations: Interventions and Interviews, 1971–2001*, edited and translated by Elizabeth Rottenberg, Stanford UP, 2002, pp. 343–71.

———. "Circumfession." *Jacques Derrida*, U of Chicago P, 1993, pp. 3–315.

———. *Clang*. Translated by Geoffrey Bennington and David Wills, U of Minnesota P, 2020.

———. *De la grammatologie*. Les Éditions de Minuit, 1967.

———. "Des Tours de Babel." *Acts of Religion*, edited by Gil Anidjar, translated by Joseph F. Graham, Routledge, 2002, pp. 102–34.

———. *Given Time: I. Counterfeit Money*. Translated by Peggy Kamuf, U of Chicago P, 1992.

———. *Life Death*. Edited by Peggy Kamuf, translated by Pascale-Anne Brault and Michael Naas, U of Chicago P, 2020.

———. *Limited Inc*. Edited by G. Graff, translated by S. Weber and J. Mehlman, Northwestern UP, 1988.

———. "Parergon." *The Truth in Painting*, translated by Geoffrey Bennington and Ian McLeod, U of Chicago P, 1987, pp. 15–148.

———. *The Politics of Friendship*. Translated by George Collins, Verso, 2005.

———. *Positions*. Translated by Alan Bass, U of Chicago P, 1972.

———. "Psyche: Invention of the Other." *Psyche: Inventions of the Other*, edited by Peggy Kamuf and Elizabeth Rottenberg, translated by Catherine Porter, vol. I, Stanford UP, 2007, pp. 1–47.

———. "Psychoanalysis Searches the States of Its Soul: The Impossible Beyond of a Sovereign Cruelty." *Without Alibi*, Stanford UP, 2002, pp. 238–80.

———. "The Rhetoric of Drugs." *Points . . . : Interviews, 1974–1994*, edited by Elisabeth Weber, translated by Michael Israel, Stanford UP, 1995, pp. 228–54.

———. "Some Statements and Truisms about Neologisms, Newisms, Postisms, Parasitisms, and Other Small Seismisms." *The States of "Theory": History, Art, and Critical Discourse*, Columbia UP, 1990, pp. 63–94.

———. "The Spatial Arts: An Interview with Jacques Derrida." *Deconstruction and the Visual Arts: Art, Media, Architecture*, edited by Peter Brunette and David Wills, Cambridge UP, 1994, pp. 9–32.

———. *Specters of Marx: The State of the Debt, the Work of Mourning, and the New International*. Translated by Peggy Kamuf, Routledge, 1994.

———. "The Time Is Out of Joint." *Deconstruction Is/In America: A New Sense of the Political*, edited by Anselm Haverkamp, NYUP, 1995, pp. 14–38, https://www.jstor.org/stable/j.ctt9qfqqx.6. JSTOR.

———. *Trace et archive, image et art*. INA éditions, 2014.

———. "White Mythology: Metaphor in the Text of Philosophy." *Margins of Philosophy*, translated by Alan Bass, U of Chicago P, 1982, pp. 207–72.

DiEuliis, Diane, et al. "Options for Synthetic DNA Order Screening, Revisited." *mSphere*, vol. 2, no. 4, Aug. 2017, https://doi.org/10.1128/mSphere.00319-17.

Domingo, Esteban, et al. "Viral Quasispecies Evolution." *Microbiology and Molecular Biology Reviews*, vol. 76, no. 2, June 2012, pp. 159–216, https://doi.org/10.1128/MMBR.05023-11.

Doolittle, W. Ford. "The Practice of Classification and the Theory of Evolution, and What the Demise of Charles Darwin's Tree of Life Hypothesis Means for Both of Them." *Philosophical Transactions of the Royal Society B: Biological Sciences*, vol. 364, no. 1527, Aug. 2009, pp. 2221–28, https://doi.org/10.1098/rstb.2009.0032.

Doolittle, W. Ford, and Eric Bapteste. "Pattern Pluralism and the Tree of Life Hypothesis." *Proceedings of the National Academy of Sciences*, vol. 104, no. 7, Feb. 2007, p. 2043, https://doi.org/10.1073/pnas.0610699104.

Dowdle, W. R. "The Principles of Disease Elimination and Eradication." *Bulletin of the World Health Organization*, vol. 76, supp. 2, 1998, pp. 22–25.

Dowdle, Walter R., and Maureen E. Birmingham. "The Biologic Principles of Poliovirus Eradication." *The Journal of Infectious Diseases*, vol. 175, supp. 1, Feb. 1997, pp. S286–92, https://doi.org/10.1093/infdis/175.Supplement_1.S286.

Duckworth, Donna H. "Who Discovered Bacteriophage?" *Bacteriological Reviews*, vol. 40, no. 4, Dec. 1976, pp. 793–802, https://doi.org/10.1128/br.40.4.793-802.1976.

Dupré, John, and Stephan Guttinger. "Viruses as Living Processes." *Studies in History and Philosophy of Science Part C: Studies in History and Philosophy of Biological and Biomedical Sciences*, vol. 59, Oct. 2016, pp. 109–16, https://doi.org/10.1016/j.shpsc.2016.02.010.

Eigen, Manfred, and Peter Schuster. "A Principle of Natural Self-Organization." *Naturwissenschaften*, vol. 64, no. 11, Nov. 1977, pp. 541–65, https://doi.org/10.1007/BF00450633.

Eisenberg, Rebecca, et al. "Symposium on Bioinformatics and Intellectual Property Law — April 27, 2001 — Molecules vs. Information: Should Patents Protect Both?" *Boston University Journal of Science and Technology Law*, vol. 8, Winter 2002, pp. 190–217.

Elgin, Ben, and Kevin Crowley. "Algae Fuel Company That Exxon Once Bankrolled Finds New Funders." *Bloomberg.com*, Mar. 2023. Business Source Ultimate.

———. "Exxon Retreats from Major Climate Effort to Make Biofuels from Algae." *Bloomberg.com*, Feb. 2023. Business Source Ultimate.

Erez, Zohar, et al. "Communication between Viruses Guides Lysis–Lysogeny Decisions." *Nature*, vol. 541, no. 7638, Jan. 2017, pp. 488–93, https://doi.org/10.1038/nature21049.

ETC Staff. "The Men and Money behind Synthia." *ETC Group*, 1 Oct. 2009, https://www.etcgroup.org/content/men-and-money-behind-synthia.

———. "Patenting Pandora's Bug: Goodbye, Dolly . . . Hello, Synthia!" *ETC Group*, 6 June 2007, https://www.etcgroup.org/content/patenting-pandora's-bug-goodbye-dollyhello-synthia.

Evans, Nicholas Greig, et al. "The Ethics of Biosafety Considerations in Gain-of-Function Research Resulting in the Creation of Potential Pandemic Pathogens." *Journal of Medical Ethics*, vol. 41, no. 11, Nov. 2015, p. 901, https://doi.org/10.1136/medethics-2014-102619.

Fan, Nana, et al. "Piglets Cloned from Induced Pluripotent Stem Cells." *Cell Research*, vol. 23, no. 1, Jan. 2013, pp. 162–66, https://doi.org/10.1038/cr.2012.176.

Flint, J., et al. *Principles of Virology*. 5th ed., vol. 1, Wiley, 2020.

Fogg, Paul C. M., et al. "Bacteriophage Lambda: A Paradigm Revisited." *Journal of Virology*, vol. 84, no. 13, 2010, pp. 6876–79, https://doi.org/10.1128/JVI.02177-09.

Forterre, Patrick. "Darwin's Goldmine Is Still Open: Variation and Selection Run the World." *Frontiers in Cellular and Infection Microbiology*, vol. 2, 2012, pp. 1–13, https://doi.org/10.3389/fcimb.2012.00106.

———. "Defining Life: The Virus Viewpoint." *Origins of Life and Evolution of Biospheres*, vol. 40, no. 2, Apr. 2010, pp. 151–60, https://doi.org/10.1007/s11084-010-9194-1.

———. "Reviewer's Report 2" in Mark Ragan, "Trees and Networks before and after Darwin." *Biology Direct*, vol. 4, no. 1, Nov. 2009, https://doi.org/10.1186/1745-6150-4-43.

———. "The Universal Tree of Life: An Update." *Frontiers in Microbiology*, vol. 6, 2015, https://doi.org/10.3389/fmicb.2015.00717.

Forterre, Patrick, et al. "Cellular Domains and Viral Lineages." *Trends in Microbiology*, vol. 22, no. 10, Oct. 2014, pp. 554–58, https://doi.org/10.1016/j.tim.2014.07.004.

Fracchia, Joseph, and R. C. Lewontin. "Does Culture Evolve?" *History and Theory*, vol. 38, no. 4, Dec. 1999, pp. 52–78, https://doi.org/10.1111/0018-2656.00104.

Freud, Sigmund. "The Aetiology of Hysteria." *The Standard Edition of the Complete Psychological Works of Sigmund Freud, Volume III (1893–1899):*

Early Psycho-Analytic Publications, The Hogarth Press and the Institute of Psycho-analysis, 1896, pp. 187–221. PEP-Web.

Fritsch, Matthias, et al., editors. *Eco-Deconstruction: Derrida and Environmental Philosophy*. Fordham UP, 2018.

———. "Introduction." *Eco-Deconstruction: Derrida and Environmental Philosophy*, edited by Matthias Fritsch et al., Fordham UP, 2018, pp. 1–28.

Fulka, Josef Jr, et al. "Nucleus Transfer in Mammals: Noninvasive Approaches for the Preparation of Cytoplasts." *Trends in Biotechnology*, vol. 22, no. 6, June 2004, pp. 279–83, https://doi.org/10.1016/j.tibtech.2004.04.002.

Fuller, Jonathan. "Models versus Evidence." *Thinking in a Pandemic: The Crisis of Science and Policy in the Age of COVID-19*, Boston Review and Verso Books, 2020, pp. 76–89.

Funnell, Barbara E., and Gregory J. Phillips. "Preface." *Plasmid Biology*, edited by Barbara E. Funnell and Gregory J. Phillips, 2004, p. xi, https://doi.org/10.1128/9781555817732.fmatter.

Gamble, Christopher N., et al. "What is New Materialism?" *Angelaki*, vol. 24, no. 6, Nov. 2019, pp. 111–34, https://doi.org/10.1080/0969725X.2019.1684704.

Gauchat, Gordon. "The Political Context of Science in the United States: Public Acceptance of Evidence-Based Policy and Science Funding." *Social Forces*, vol. 94, no. 2, Dec. 2015, pp. 723–46, https://doi.org/10.1093/sf/sov040.

Georgiades, Kalliopi, and Didier Raoult. "How Microbiology Helps Define the Rhizome of Life." *Frontiers in Cellular and Infection Microbiology*, vol. 2, 2012, p. 60, https://doi.org/10.3389/fcimb.2012.00060.

Gibson, Daniel G., et al. "Complete Chemical Synthesis, Assembly, and Cloning of a Mycoplasma Genitalium Genome." *Science*, vol. 319, no. 5867, Feb. 2008, pp. 1215–20, https://doi.org/10.1126/science.1151721.

Gibson, Daniel G., et al. "Creation of a Bacterial Cell Controlled by a Chemically Synthesized Genome." *Science*, vol. 329, no. 5987, May 2010, pp. 52–56, https://doi.org/10.1126/science.1190719.

Gibson, Daniel G., et al. Methods for in vitro joining and combinatorial assembly of nucleic acid molecules. US 8968999 B2, United States Patent and Trademark Office, 3 March 2015. *Google Patents,* https://patents.google.com/patent/US8968999B2/.

Gilbert, Scott F. "The Genome in Its Ecological Context." *Annals of the New York Academy of Sciences*, vol. 981, no. 1, Dec. 2002, pp. 202–18, https://doi.org/10.1111/j.1749-6632.2002.tb04919.x.

Gilbert, Scott F., and David Epel. *Ecological Developmental Biology: The Environmental Regulation of Development, Health, and Evolution*. Sinauer Associates, 2015.

Gilbert, Scott F., et al. "A Symbiotic View of Life: We Have Never Been Individuals." *The Quarterly Review of Biology*, vol. 87, no. 4, Dec. 2012, pp. 325–41, https://doi.org/10.1086/668166.

Giuntini, J., et al. "Étude de Quelques Bactériophages typhiques Vi. Morphologie des Corpuscules au Microscope Électronique Aspect des plages et thermosensibilité." *Annales de l'Institut Pasteur*, vol. 84, 1953, pp. 787–91.

Giuntini, J., et al. "Images électroniques de quelques bactériophages et détermination de leur taille." *Annales de l'institut Pasteur*, vol. 73, 1947, pp. 579–81.

Glass, John I., et al. Installation of genomes or partial genomes into cells or cell-like systems. US 9434974 B2, United States Patent and Trademark Office, 6 September 2016. *Google Patents,* https://patents.google.com/patent/US9434974B2/.

Glass, John I., et al. Minimal bacterial genome. US 20070122826 A1, United States Patent and Trademark Office, 31 May 2007. *Google Patents,* https://patents.google.com/patent/US20070122826A1/.

Glass, John I., et al. "Minimal Cells — Real and Imagined." *Cold Spring Harbor Perspectives in Biology*, vol. 9, no. 12, Dec. 2017, https://doi.org/10.1101/cshperspect.a023861.

Gogarten, J. Peter, et al. "Prokaryotic Evolution in Light of Gene Transfer." *Molecular Biology and Evolution*, vol. 19, Jan. 2003, pp. 2226–38, https://doi.org/10.1093/oxfordjournals.molbev.a004046.

Gómez-Tatay, Lucía, and José M. Hernández-Andreu. "Biosafety and Biosecurity in Synthetic Biology: A Review." *Critical Reviews in Environmental Science and Technology*, vol. 49, no. 17, Sept. 2019, pp. 1587–621, https://doi.org/10.1080/10643389.2019.1579628.

Gordillo Altamirano, Fernando L., and Jeremy J. Barr. "Phage Therapy in the Postantibiotic Era." *Clinical Microbiology Reviews*, vol. 32, no. 2, Mar. 2019, https://doi.org/10.1128/CMR.00066-18.

Gostin, Lawrence O. "Global Polio Eradication: Espionage, Disinformation, and the Politics of Vaccination." *The Milbank Quarterly*, vol. 92, no. 3, Sept. 2014, pp. 413–17, https://doi.org/10.1111/1468-0009.12065.

Gould, Stephen Jay. "Evolution and the Triumph of Homology, or Why History Matters." *American Scientist*, vol. 74, no. 1, 1986, pp. 60–69. JSTOR.

Goulian, M., et al. "Enzymatic Synthesis of DNA, XXIV. Synthesis of Infectious Phage Phi-X174 DNA." *PNAS*, vol. 58, no. 6, Dec. 1967, pp. 2321–28, https://doi.org/10.1073/pnas.58.6.2321.

Gouveia, Chantel, et al. "Lessons Learned from Somatic Cell Nuclear Transfer." *International Journal of Molecular Sciences*, vol. 21, no. 7, 2020, https://doi.org/10.3390/ijms21072314.

Grasis, Juris A. "The Intra-Dependence of Viruses and the Holobiont." *Frontiers in Immunology*, vol. 8, 2017, p. 1501, https://doi.org/10.3389/fimmu.2017.01501.

Groenke, Nicole, et al. "Mechanism of Virus Attenuation by Codon Pair Deoptimization." *Cell Reports*, vol. 31, no. 4, Apr. 2020, https://doi.org/10.1016/j.celrep.2020.107586.

Gruenberg, Benjamin C. "Artificial Life." *Scientific American*, vol. 105, no. 13, 1911, pp. 272–86. JSTOR.

Haig, David. *From Darwin to Derrida: Selfish Genes, Social Selves, and the Meanings of Life*. MIT Press, 2020.

Halary, Sébastien, et al. "Network Analyses Structure Genetic Diversity in Independent Genetic Worlds." *Proceedings of the National Academy of Sciences*, vol. 107, no. 1, Jan. 2010, pp. 127–32, https://doi.org/10.1073/pnas.0908978107.

Haldane, John Burdon Sanderson. "The Origin of Life." *Rationalist Annual*, vol. 148, 1929, pp. 3–10.

Hall, Brian K. "Introduction." *Homology: The Hierarchical Basis of Comparative Biology*, Academic Press, 1994, pp. 1–20.

Haraway, Donna. "Anthropocene, Capitalocene, Plantationocene, Chthulucene: Making Kin." *Environmental Humanities*, vol. 6, no. 1, May 2015, pp. 159–65, https://doi.org/10.1215/22011919-3615934.

———. "A Cyborg Manifesto: Science, Technology, and Socialist-Feminism in the Late Twentieth Century." *Simians, Cyborgs, and Women: The Reinvention of Nature*, Routledge, 1991, pp. 149–82.

———. *Modest_Witness@Second_Millennium.FemaleMan_Meets_OncoMouse: Feminism and Technoscience*. 2nd ed., Routledge, 2018.

———. *Staying with the Trouble: Making Kin in the Chthulucene*. Duke UP, 2016.

———. *When Species Meet*. U of Minnesota P, 2008.

Harris, Hugh M. B., and Colin Hill. "A Place for Viruses on the Tree of Life." *Frontiers in Microbiology*, vol. 11, 2021, https://doi.org/10.3389/fmicb.2020.604048.

Hatch, Anthony Ryan. "Two Meditations in Coronatime." *SKAT: The Science Knowledge and Technology Section of the American Sociological Association*, 22 May 2020, https://asaskat.com/2020/05/22/two-meditations-in-coronatime/.

Hawkins, David. *The Language of Nature: An Essay in the Philosophy of Science*. San Francisco, W. H. Freeman, 1964.

Hayes, W. "Recombination in *Bact. coli K* 12: Unidirectional Transfer of Genetic Material." *Nature*, vol. 169, no. 4290, Jan. 1952, pp. 118–19, https://doi.org/10.1038/169118b0.

———. "What Are Episomes and Plasmids?" *Bacterial Episomes and Plasmids*, edited by G. E. W. Wolstenholme and Maeve O'Connor, J & A Churchill, 1969, pp. 4–11.

Hayles, N. Katherine. *How We Became Posthuman: Virtual Bodies in Cybernetics, Literature, and Informatics*. U of Chicago P, 1999.

Heidegger, Martin. *On the Way to Language*. Translated by Peter D. Hertz, HarperCollins, 1982.

Heise, Ursula K. *Imagining Extinction: The Cultural Meanings of Endangered Species*. U of Chicago P, 2016.

Hickey, Donal A. "Selfish DNA: A Sexually-Transmitted Nuclear Parasite." *Genetics*, vol. 101, no. 3–4, July 1982, pp. 519–31.

Hirsch, Jesse. "The Life and Times of Dolly the Cloned Sheep." *Modern Farmer*, 10 Dec. 2013, https://modernfarmer.com/2013/12/life-times-dolly-clone/.

Hobbs, Zack, and Stephen T. Abedon. "Diversity of Phage Infection Types and Associated Terminology: The Problem with 'Lytic or Lysogenic.'" *FEMS Microbiology Letters*, vol. 363, no. 7, Apr. 2016, https://doi.org/10.1093/femsle/fnw047.

Holmes, Edward C. *The Evolution and Emergence of RNA Viruses*. Oxford UP, 2009.

Hoose, Alex, et al. "DNA Synthesis Technologies to Close the Gene Writing Gap." *Nature Reviews Chemistry*, vol. 7, no. 3, Mar. 2023, pp. 144–61, https://doi.org/10.1038/s41570-022-00456-9.

Horowitz, N. H. "The Evolution of Biochemical Syntheses — Retrospect and Prospect." *Evolving Genes and Proteins: A Symposium*, edited by V. Bryson and H. J. Vogel, Academic Press, 1965.

Horzinek, Marian C. "The Birth of Virology." *Antonie van Leeuwenhoek*, vol. 71, no. 1, Feb. 1997, pp. 15–20, https://doi.org/10.1023/A:1000197505492.

———. "Editorial." *Archives of Virology*, vol. 140, no. 4, Apr. 1995, pp. 807–08, https://doi.org/10.1007/BF01309969.

Hossain, Tahmina, et al. "Antibiotic Tolerance, Persistence, and Resistance of the Evolved Minimal Cell, Mycoplasma Mycoides JCVI-Syn3B." *iScience*, vol. 24, no. 5, May 2021, https://doi.org/10.1016/j.isci.2021.102391.

"How Scientists Made 'Artificial Life.'" *BBC News*, http://news.bbc.co.uk/2/hi/science/nature/8695992.stm. Accessed 8 Jan. 2020.

"How the CIA's Fake Vaccination Campaign Endangers Us All." *Scientific American*, https://doi.org/10.1038/scientificamerican0513-12. Accessed 27 May 2021.

Howell, Katie. "Exxon Sinks $600M into Algae-Based Biofuels in Major Strategy Shift." *NYTimes*, 14 July 2009, https://archive.nytimes.com/www.nytimes.

com/gwire/2009/07/14/14greenwire-exxon-sinks-600m-into-algae-based-biofuels-in-33562.html.

Hughes, Sally Smith. *The Virus: A History of the Concept*. Heinemann Educational Books, 1977.

Hutchison, Clyde A., et al. "Design and Synthesis of a Minimal Bacterial Genome." *Science*, vol. 351, no. 6280, 2016, https://doi.org/10.1126/science.aad6253.

Hutchison, Clyde A., et al. "Global Transposon Mutagenesis and a Minimal Mycoplasma Genome." *Science*, vol. 286, no. 5447, Dec. 1999, pp. 2165–69, https://doi.org/10.1126/science.286.5447.2165.

Ingraham, Christopher. "An Alarming Number of Scientific Papers Contain Excel Errors." *Washington Post*, 26 Aug. 2016. *www.washingtonpost.com*, https://www.washingtonpost.com/news/wonk/wp/2016/08/26/an-alarming-number-of-scientific-papers-contain-excel-errors/.

IUCN SSC. *IUCN SSC Guiding Principles on Creating Proxies of Extinct Species for Conservation Benefit*. Version 1.0, IUCN Species Survival Commission, 2016.

Ivanowski, Dmitrii. "Concerning the Mosaic Disease of the Tobacco Plant, in: Early Papers on Tobacco Mosaic and Infectious Variegation." *Phytopathological Classics*, vol. 7, 1942, pp. 27–30, https://doi.org/10.1094/9780890545225.001.

Jablonka, Eva, and Marion J. Lamb. *Evolution in Four Dimensions: Genetic, Epigenetic, Behavioral, and Symbolic Variation in the History of Life*. MIT Press, 2014.

Jackson, Zakiyyah Iman. *Becoming Human: Matter and Meaning in an Antiblack World*. NYUP, 2020.

Jacob, François. *Genetic Control of Viral Functions*. 1958, https://wellcomecollection.org/works/e895knvt. Sydney Brenner Collection, Wellcome Collection, SB/2/3/91.

———. *The Logic of Life: A History of Heredity*. Translated by Betty E. Spillmann, Princeton UP, 1993.

Jacob, François, and Ellie L. Wollman. *Sexuality and the Genetics of Bacteria*. Academic Press, 1961.

Jacob, François, et al. "Definition of Some Terms Relative to Lysogeny." *Annales de l'Institut Pasteur*, vol. 84, no. 1, 1953, pp. 222–24.

Johnston, John. *The Allure of Machinic Life: Cybernetics, Artificial Life, and the New AI*. MIT Press, 2008.

Kahl, Linda J., and Drew Endy. "A Survey of Enabling Technologies in Synthetic Biology." *Journal of Biological Engineering*, vol. 7, no. 1, May 2013, p. 13, https://doi.org/10.1186/1754-1611-7-13.

Kamuf, Peggy. "Derrida and Gender: The Other Sexual Difference." *Jacques Derrida and the Humanities*, Cambridge UP, 2002.

Kant, Immanuel. *Critique of Judgment*. Translated by Werner Pluhar, Hackett, 1987.

Karera, Axelle. "Blackness and the Pitfalls of Anthropocene Ethics." *Critical Philosophy of Race*, vol. 7, no. 1, Jan. 2019, pp. 32–56, https://doi.org/10.5325/critphilrace.7.1.0032.

Kay, Lily E. "W. M. Stanley's Crystallization of the Tobacco Mosaic Virus, 1930–1940." *Isis*, vol. 77, no. 3, Sept. 1986, pp. 450–72, https://doi.org/10.1086/354205.

Keller, Evelyn Fox. *The Century of the Gene*. Harvard UP, 2002.

———. "Knowing as Making, Making as Knowing: The Many Lives of Synthetic Biology." *Biological Theory*, vol. 4, no. 4, Dec. 2009, pp. 333–39, https://doi.org/10.1162/BIOT_a_00005.

———. *Making Sense of Life: Explaining Biological Development with Models, Metaphors, and Machines*. Harvard UP, 2002.

———. *Secrets of Life, Secrets of Death: Essays on Language, Gender, and Science*. Routledge, 1992.

———. "What Does Synthetic Biology Have to Do with Biology?" *BioSocieties*, vol. 4, no. 2–3, 2009, pp. 291–302. Cambridge Core, *Cambridge UP*, https://doi.org/10.1017/S1745855209990123.

Keller, Evelyn Fox, and Jeremy Ahouse. "Writing and Reading about Dolly." *BioEssays*, vol. 19, no. 8, Aug. 1997, pp. 740–42, https://doi.org/10.1002/bies.950190816.

Kennedy, Jonathan. "How Drone Strikes and a Fake Vaccination Program Have Inhibited Polio Eradication in Pakistan: An Analysis of National Level Data." *International Journal of Health Services*, vol. 47, no. 4, Aug. 2017, pp. 807–25, https://doi.org/10.1177/0020731417722888.

Kimmerer, Robin Wall. *Braiding Sweetgrass: Indigenous Wisdom, Scientific Knowledge and the Teachings of Plants*. Milkweed Editions, 2013.

Kirby, Vicki, et al. "How Do We Do Biodeconstruction?" *Postmodern Culture*, vol. 28, no. 3, 2018. *Project MUSE*, https://doi.org/10.1353/pmc.2018.0021.

———. *Quantum Anthropologies*. Duke UP, 2011.

Kjær, Kurt H., et al. "A 2-Million-Year-Old Ecosystem in Greenland Uncovered by Environmental DNA." *Nature*, vol. 612, no. 7939, Dec. 2022, pp. 283–91, https://doi.org/10.1038/s41586-022-05453-y.

Kjær, Kurt H., and Eske Willerslev. "DNA Reveals That Mastodons Roamed a Forested Greenland Two Million Years Ago." *Nature*, Dec. 2022. *www.nature.com*, https://doi.org/10.1038/d41586-022-03626-3.

Kohler, Robert E. "Drosophila: A Life in the Laboratory." *Journal of the History of Biology*, vol. 26, no. 2, 1993, pp. 281–310. JSTOR.

Koonin, Eugene V. "Comparative Genomics, Minimal Gene-Sets and the Last Universal Common Ancestor." *Nature Reviews Microbiology*, vol. 1, no. 2, Nov. 2003, pp. 127–36, https://doi.org/10.1038/nrmicro751.

———. *The Logic of Chance: The Nature and Origin of Biological Evolution*. Pearson Education, 2011.

Köpke, Michael. "Redesigning CO_2 Fixation." *Nature Synthesis*, vol. 1, no. 8, Aug. 2022, pp. 584–85, https://doi.org/10.1038/s44160-022-00131-3.

Kornbluh, Anna. "We Didn't Start the Fire: Death Drive against Ecocide." *Parapraxis*, Dec. 2023, https://www.parapraxismagazine.com/articles/we-didnt-start-the-fire.

Kriss, A. E. "Polymerity in the Structural Organization of Bacteriophages." *Verhandlungen Band II / Biologisch-Medizinischer Teil*, edited by W. Bargmann et al., Springer Berlin Heidelberg, 1960, pp. 621–25.

Kriss, A. E., and A. S. Tikhonenko. "Structure of the Corpuscle of Bacteriophage." *Doklady Akademii nauk SSSR*, vol. 86, no. 2, Sept. 1952, pp. 421–23.

Kritz, Fran. "Russian Lab Explosion Raises Question: Should Smallpox Virus Be Kept or Destroyed?" *NPR.Org*, 19 Sept. 2019, https://www.npr.org/sections/goatsandsoda/2019/09/19/762013515/russian-lab-explosion-raises-question-should-smallpox-virus-be-kept-or-destroyed.

Kuchment, Anna. *The Forgotten Cure: The Past and Future of Phage Therapy*. Springer, 2012.

Kuhn, Thomas S. "Commensurability, Comparability, Communicability." *The Road since Structure: Philosophical Essays, 1970–1993, with an Autobiographical Interview*, edited by J. Conant and J. Haugeland, U of Chicago P, 2000, pp. 33–57.

———. *The Structure of Scientific Revolutions*. 4th ed., U of Chicago P, 2012.

Kunin, Victor, et al. "The Net of Life: Reconstructing the Microbial Phylogenetic Network." *Genome Research*, vol. 15, no. 7, July 2005, pp. 954–59, https://doi.org/10.1101/gr.3666505.

Kwok, Roberta. "Five Hard Truths for Synthetic Biology." *Nature*, vol. 463, no. 7279, Jan. 2010, pp. 288–90, https://doi.org/10.1038/463288a.

———. "Genomics: DNA's Master Craftsmen." *Nature*, vol. 468, no. 7320, Nov. 2010, pp. 22–25, https://doi.org/10.1038/468022a.

La Scola, Bernard, et al. "A Giant Virus in Amoebae." *Science*, vol. 299, no. 5615, Mar. 2003, p. 2033, https://doi.org/10.1126/science.1081867.

La Scola, Bernard, et al. "The Virophage as a Unique Parasite of the Giant Mimivirus." *Nature*, vol. 455, no. 7209, Sept. 2008, pp. 100–04, https://doi.org/10.1038/nature07218.

Lagesen, Karin, et al. "Genome Update: The 1000th Genome – a Cautionary Tale." *Microbiology*, vol. 156, no. 3, Microbiology Society, 2010, pp. 603–08, https://doi.org/10.1099/mic.0.038257-0.

Lakoff, Andrew. *Unprepared: Global Health in a Time of Emergency*. U of California P, 2017.

Lamey, Andy. "The De-Extinction of the Woolly Mammoth Is a Legal and Regulatory Nightmare." *The New Republic*, 15 Dec. 2022. *The New Republic*, https://newrepublic.com/article/169528/woolly-mammoth-alaska-colossal-thiel.

Land, Miriam, et al. "Insights from 20 Years of Bacterial Genome Sequencing." *Functional & Integrative Genomics*, vol. 15, no. 2, Mar. 2015, pp. 141–61, https://doi.org/10.1007/s10142-015-0433-4.

Lander, Eric S., et al. "Initial Sequencing and Analysis of the Human Genome." *Nature*, vol. 409, no. 6822, Feb. 2001, pp. 860–921, https://doi.org/10.1038/35057062.

Langton, Christopher G. "Artificial Life." *Artificial Life: The Proceedings of an Interdisciplinary Workshop on the Synthesis and Simulation of Living Systems, Held September, 1987, in Los Alamos, New Mexico*, edited by Christopher G. Langton, Addison-Wesley Pub. Co., Advanced Book Program, 1989, pp. 1–48.

Lartigue, Carole, et al. "Genome Transplantation in Bacteria: Changing One Species to Another." *Science*, vol. 317, no. 5838, Aug. 2007, pp. 632–38, https://doi.org/10.1126/science.1144622.

Latour, Bruno. *Facing Gaia: Eight Lectures on the New Climatic Regime*. Translated by Catherine Porter, Polity Press, 2017.

———. *The Pasteurization of France*. Translated by A. Sheridan and J. Law, Harvard UP, 1988.

Lecoq, Hervé. "Découverte du premier virus, le virus de la mosaïque du tabac : 1892 ou 1898 ?" *Comptes Rendus de l'Académie des Sciences - Series III - Sciences de la Vie*, vol. 324, no. 10, Oct. 2001, pp. 929–33, https://doi.org/10.1016/S0764-4469(01)01368-3.

Lederberg, Joshua. "Cell Genetics and Hereditary Symbiosis." *Physiological Reviews*, vol. 32, no. 4, Oct. 1952, pp. 403–30, https://doi.org/10.1152/physrev.1952.32.4.403.

Leduc, Stéphane. *La Biologie synthétique*. A. Poinat, 1912.

———. *The Mechanism of Life*. Translated by W. Deane Butcher, William Heinemann, 1914.

Legendre, Matthieu, et al. "Thirty-Thousand-Year-Old Distant Relative of Giant Icosahedral DNA Viruses with a Pandoravirus Morphology." *PNAS*, vol. 111, no. 11, Mar. 2014, p. 4274, https://doi.org/10.1073/pnas.1320670111.

Lehman, Joel, et al. "The Surprising Creativity of Digital Evolution: A Collection of Anecdotes from the Evolutionary Computation and Artificial Life Research Communities." *Artificial Life*, vol. 26, no. 2, May 2020, pp. 274–306, https://doi.org/10.1162/artl_a_00319.

Levins, Richard, and Richard C. Lewontin. *The Dialectical Biologist*. Harvard UP, 1985.

Lewontin, Richard C. *The Triple Helix: Gene, Organism, and Environment*. Harvard UP, 2000.

Lipsitch, Marc, and Alison P. Galvani. "Ethical Alternatives to Experiments with Novel Potential Pandemic Pathogens." *PLOS Medicine*, vol. 11, no. 5, May 2014, https://doi.org/10.1371/journal.pmed.1001646.

Loeb, Jacques. *The Mechanistic Conception of Life*. U of Chicago P, 1912.

López-García, Purificación, et al. "Metagenome-Derived Virus-Microbe Ratios across Ecosystems." *The ISME Journal*, vol. 17, no. 10, Oct. 2023, pp. 1552–63, https://doi.org/10.1038/s41396-023-01431-y.

Lukjancenko, Oksana, et al. "Comparison of 61 Sequenced Escherichia Coli Genomes." *Microbial Ecology*, vol. 60, no. 4, Nov. 2010, pp. 708–20, https://doi.org/10.1007/s00248-010-9717-3.

Luria, S. E. "Bacteriophage: An Essay on Virus Reproduction." *Science*, vol. 111, no. 2889, May 1950, p. 507, https://doi.org/10.1126/science.111.2889.507.

———. "Electron Microscope Studies of Bacterial Viruses." *Journal of Bacteriology*, vol. 46, no. 1, July 1943, pp. 57–77.

Lustig, A., and A. J. Levine. "One Hundred Years of Virology." *Journal of Virology*, vol. 66, no. 8, Aug. 1992, p. 4629.

Lwoff, André. "The Concept of Virus." *Microbiology*, vol. 17, no. 2, Microbiology Society, 1957, pp. 239–53, https://doi.org/10.1099/00221287-17-2-239.

———. "Lysogeny." *Bacteriology Reviews*, vol. 17, no. 4, Dec. 1953, pp. 269–337.

Lynes, Philippe. *Futures of Life Death on Earth: Derrida's General Ecology*. Rowman & Littlefield International, 2018.

Madrigal, Alexis. "Scientists Build First Man-Made Genome; Synthetic Life Comes Next." *Wired*, 24 Jan. 2008. *www.wired.com*, https://www.wired.com/2008/01/scientists-build-first-man-made-genome-synthetic-life-comes-next/.

Malabou, Catherine. *Plasticity: The Promise of Explosion*. Edited by Tyler M. Williams, Edinburgh UP, 2022.

Malin, Katarzyna, et al. "The Many Problems of Somatic Cell Nuclear Transfer in Reproductive Cloning of Mammals." *Theriogenology*, vol. 189, Sept. 2022, pp. 246–54, https://doi.org/10.1016/j.theriogenology.2022.06.030.

Malm, Andreas. *Corona, Climate, Chronic Emergency: War Communism in the Twenty-First Century*. Verso, 2020.

Malm, Andreas and The Zetkin Collective. *White Skin, Black Fuel: On the Danger of Fossil Fascism*. Verso, 2021.

Marder, Elissa. "Insex." *Parallax*, vol. 25, no. 2, Apr. 2019, pp. 228–39, https://doi.org/10.1080/13534645.2019.1607229.

———. *The Mother in the Age of Mechanical Reproduction*. Fordham UP, 2012.

———. "Pandora's Fireworks; or, Questions Concerning Femininity, Technology, and the Limits of the Human." *Philosophy & Rhetoric*, vol. 47, no. 4, 2014, pp. 386–99, https://doi.org/10.5325/philrhet.47.4.0386. JSTOR.

Margulis, Lynn, and Dorion Sagan. *Acquiring Genomes: A Theory of the Origins of Species*. Basic Books, 2002.

Martinez-Bravo, Monica, and Andreas Stegmann. "In Vaccines We Trust? The Effects of the CIA's Vaccine Ruse on Immunization in Pakistan." *Journal of the European Economic Association*, vol. 20, no. 1, May 2021, pp. 150–86, https://doi.org/10.1093/jeea/jvab018.

Mastrogiovanni, Armando M. "Biosignature, Technosignature, Event: Deconstruction, Astrobiology, and the Search for a Wholly Other Origin." *Derrida Today*, vol. 16, no. 2, Nov. 2023, pp. 114–28, https://doi.org/10.3366/drt.2023.0312.

Matthews, Christopher M. "Exxon Sees Green Gold in Algae-Based Fuels. Skeptics See Greenwashing." *Wall Street Journal*, 3 Oct. 2021. *www.wsj.com*, https://www.wsj.com/articles/exxon-sees-green-gold-in-algae-based-fuels-skeptics-see-greenwashing-11633258802.

Mayer, Adolf. "Concerning the Mosaic Disease of Tobacco." *Phytopathological Classics*, vol. 7, 1942, pp. 11–24, https://doi.org/10.1094/9780890545225.001.

Mayer, Jens, et al. "A Revised Nomenclature for Transcribed Human Endogenous Retroviral Loci." *Mobile DNA*, vol. 2, no. 1, May 2011, p. 7, https://doi.org/10.1186/1759-8753-2-7.

Mayr, Ernst. "Cause and Effect in Biology." *Science*, vol. 134, no. 3489, Nov. 1961, pp. 1501–06.

———. *The Growth of Biological Thought: Diversity, Evolution, and Inheritance*. Belknap Press, 1982.

McCance, Dawne. *The Reproduction of Life Death: Derrida's La Vie La Mort*. Fordham UP, 2019.

McClintock, Barbara. "The Significance of Responses of the Genome to Challenge." *Science*, vol. 226, no. 4676, 1984, pp. 792–801.

McLennan, Alison. *Regulation of Synthetic Biology: BioBricks, Biopunks and Bioentrepreneurs*. Edward Elgar Publishing, 2018, https://doi.org/10.4337/9781785369445.

McNeil Jr, Donald G. "They Crawl, They Bite, They Baffle Scientists." *The New York Times*, 30 Aug. 2010. *NYTimes.com*, https://www.nytimes.com/2010/08/31/science/31bedbug.html.

Meine, Curt. "De-Extinction and the Community of Being." *Hastings Center Report*, vol. 47, no. S2, July 2017, pp. S9–17, https://doi.org/10.1002/hast.746.

Mentz, Steve. *Break Up the Anthropocene*. U of Minnesota P, 2019.

Merchant, Emily Klancher. *Building the Population Bomb*. Oxford UP, 2021.

Mercier, Thomas Clément. "Resisting the Present: Biopower in the Face of the Event (Some Notes on Monstrous Lives)." *CR: The New Centennial Review*, vol. 19, no. 3, 2019, pp. 99–127.

———. "Uses of 'the Pluriverse': Cosmos, Interrupted — or the Others of Humanities." *Ostium*, vol. 15, no. 2, 2019, https://ostium.sk/language/sk/uses-of-the-pluriverse-cosmos-interrupted-or-the-others-of-humanities/.

Merhej, Vicky, and Didier Raoult. "Rhizome of Life, Catastrophes, Sequence Exchanges, Gene Creations, and Giant Viruses: How Microbial Genomics Challenges Darwin." *Frontiers in Cellular and Infection Microbiology*, vol. 2, 2012, p. 113, https://doi.org/10.3389/fcimb.2012.00113.

Merton, Robert K. *The Sociology of Science: Theoretical and Empirical Investigations*. Edited by Norman W. Storer, U of Chicago P, 1973.

Meyer, Hermann, et al. "Smallpox in the Post-Eradication Era." *Viruses*, vol. 12, no. 2, 2020, https://doi.org/10.3390/v12020138.

Mi, Sha, et al. "Syncytin Is a Captive Retroviral Envelope Protein Involved in Human Placental Morphogenesis." *Nature*, vol. 403, no. 6771, Feb. 2000, pp. 785–89, https://doi.org/10.1038/35001608.

Miller, M. H. *The Philosopher in Plato's Statesman*. Springer Netherlands, 2012.

Minteer, Ben A. *The Fall of the Wild: Extinction, De-Extinction, and the Ethics of Conservation*. Columbia UP, 2018.

Mirowski, Philip. "The Future(s) of Open Science." *Social Studies of Science*, vol. 48, no. 2, Apr. 2018, pp. 171–203, https://doi.org/10.1177/0306312718772086.

Mitman, Gregg. "Reflections on the Plantationocene: A Conversation with Donna Haraway and Anna Tsing." *Edge Effects*, 18 June 2019, https://edgeeffects.net/haraway-tsing-plantationocene/.

Moger-Reischer, R. Z., et al. "Evolution of a Minimal Cell." *Nature*, vol. 620, no. 7972, Aug. 2023, pp. 122–27, https://doi.org/10.1038/s41586-023-06288-x.

Monsanto. "Monsanto Acquires Select Assets of Agradis, Inc. to Support Work in Agricultural Biologicals." *PR Newswire*, 30 Jan. 2013, https://www.prnewswire.com/news-releases/monsanto-acquires-select-assets-of-agradis-inc-to-support-work-in-agricultural-biologicals-189005461.html.

Moore, Jason W., editor. *Anthropocene Or Capitalocene?: Nature, History, and the Crisis of Capitalism*. PM Press, 2016.

———. *Capitalism in the Web of Life: Ecology and the Accumulation of Capital*. Verso Books, 2015.

Moralee, Dennis. "Memory Protection: Avoiding the Cancer of Crashed Systems." *Electronics & Power*, vol. 28, no. 3, Mar. 1982, pp. 256–59, https://doi.org/10.1049/ep.1982.0110.

Morens, David M., et al. "Escaping Pandora's Box—Another Novel Coronavirus." *New England Journal of Medicine*, vol. 382, no. 14, Apr. 2020, pp. 1293–95, https://doi.org/10.1056/NEJMp2002106.

Morton, Timothy. *Dark Ecology: For a Logic of Future Coexistence*. Columbia UP, 2016.

Muller, H. J. "Variation Due to Change in the Individual Gene." *The American Naturalist*, vol. 56, no. 642, Jan. 1922, pp. 32–50, https://doi.org/10.1086/279846.

Naas, Michael. *Plato and the Invention of Life*. Fordham UP, 2018.

Nielsen, Alec A. K., et al. "Genetic Circuit Design Automation." *Science*, vol. 352, no. 6281, Apr. 2016, https://doi.org/10.1126/science.aac7341.

Nielsen, Jens, and Jay D. Keasling. "Synergies between Synthetic Biology and Metabolic Engineering." *Nature Biotechnology*, vol. 29, no. 8, Aug. 2011, pp. 693–95, https://doi.org/10.1038/nbt.1937.

Nobrega, Franklin L., et al. "Targeting Mechanisms of Tailed Bacteriophages." *Nature Reviews Microbiology*, vol. 16, no. 12, Dec. 2018, pp. 760–73, https://doi.org/10.1038/s41579-018-0070-8.

Nocard, (Edmond), et al. "Le microbe de la péripneumonie." *Annales de l'Institut Pasteur*, vol. 12, 1898, pp. 240–62. Biodiversity Heritage Library.

Nocard, Edmond, et al. "The Microbe of Pleuropneumonia." *Reviews of Infectious Diseases*, vol. 12, no. 2, 1990, pp. 354–58. JSTOR.

Novak, Ben J. "De-Extinction." *Genes*, vol. 9, no. 11, 2018, https://doi.org/10.3390/genes9110548.

Nyholm, Spencer V., and Margaret McFall-Ngai. "The Winnowing: Establishing the Squid–Vibrio Symbiosis." *Nature Reviews Microbiology*, vol. 2, no. 8, Aug. 2004, pp. 632–42, https://doi.org/10.1038/nrmicro957.

Ogburn, William F., and Dorothy Thomas. "Are Inventions Inevitable? A Note on Social Evolution." *Political Science Quarterly*, vol. 37, no. 1, 1922, pp. 83–98, https://doi.org/10.2307/2142320. JSTOR.

Oikkonen, Venla. "Mitochondrial Eve and the Affective Politics of Scientific Technologies." *Population Genetics and Belonging: A Cultural Analysis of Genetic Ancestry*, edited by Venla Oikkonen, Springer International Publishing, 2018, pp. 23–71, https://doi.org/10.1007/978-3-319-62881-3_2.

Olby, Robert C. *The Path to the Double Helix*. U of Washington P, 1974.

Oyama, Susan. *The Ontogeny of Information: Developmental Systems and Evolution*. Duke UP, 2000.

Panchen, Alec L. *Classification, Evolution, and the Nature of Biology*. Cambridge UP, 1992.

———. "Richard Owen and the Concept of Homology." *Homology: The Hierarchical Basis of Comparative Biology*, Academic Press, 1994, pp. 21–62.

Pasek, Anne. "Carbon Vitalism: Life and the Body in Climate Denial." *Environmental Humanities*, vol. 13, no. 1, May 2021, pp. 1–20, https://doi.org/10.1215/22011919-8867175.

Pasteur, Louis. "M. Pasteur on Hydrophobia." *Microbes and the Microbe Killer*, by William Radam, 1890, pp. 287–98, https://books.google.com/books?id=AHxIAAAAYAAJ.

Pelletier, James F., et al. "Genetic Requirements for Cell Division in a Genomically Minimal Cell." *Cell*, vol. 184, no. 9, Apr. 2021, pp. 2430–40, https://doi.org/10.1016/j.cell.2021.03.008.

Penny, David. "Darwin's Theory of Descent with Modification, versus the Biblical Tree of Life." *PLOS Biology*, vol. 9, no. 7, July 2011, https://doi.org/10.1371/journal.pbio.1001096.

Penso, Giuseppe. "Cycle of Phage Development within the Bacterial Cell." *Protoplasma*, vol. 45, no. 2, June 1955, pp. 251–63, https://doi.org/10.1007/BF01253412.

Philippe, Nadège, et al. "Pandoraviruses: Amoeba Viruses with Genomes Up to 2.5 Mb Reaching That of Parasitic Eukaryotes." *Science*, vol. 341, no. 6143, July 2013, p. 281, https://doi.org/10.1126/science.1239181.

Piper, Kelsey. "The Next Deadly Pathogen Could Come from a Rogue Scientist. Here's How We Can Prevent That." *Vox*, 11 Feb. 2020, https://www.vox.com/future-perfect/2020/2/11/21076585/dna-synthesis-assembly-viruses-biosecurity.

Plato. *Lysis*. *Complete Works*. Edited by John M. Cooper and D. S. Hutchinson, translated by Stanley Lombardo, Hackett Publishing, 1997, pp. 687–707.

———. *Republic. Complete Works*. Edited by John M. Cooper and D. S. Hutchinson, translated by G. M. A. Grube and C. D. C. Reeve, Hackett Publishing, 1997, pp. 971–1223.

———. *Statesman. Complete Works*, edited by John M. Cooper and D. S. Hutchinson, translated by C. J. Rowe, Hackett Publishing, 1997, pp. 294–358.

Pollack, Andrew. "Traces of Terror: The Science; Scientists Create a Live Polio Virus." *The New York Times*, 12 July 2002. *NYTimes.com*, https://www.nytimes.com/2002/07/12/us/traces-of-terror-the-science-scientists-create-a-live-polio-virus.html.

Porcar, Manuel, and Juli Peretó. "Are We Doing Synthetic Biology?" *Systems and Synthetic Biology*, vol. 6, no. 3, Dec. 2012, pp. 79–83, https://doi.org/10.1007/s11693-012-9101-3.

Potter, Matt. "Craig Venter's Human Genome Project on a Bluff in La Jolla: UCSD Homeboy to the Bone." *San Diego Reader*, 25 Nov. 2008, https://www.sandiegoreader.com/news/2008/nov/25/cover/.

Powell, Kendall. "How Biologists Are Creating Life-like Cells from Scratch." *Nature*, vol. 563, no. 7730, Nov. 2018, pp. 172–75. *www.nature.com*, https://doi.org/10.1038/d41586-018-07289-x.

Rabinow, Paul, and Talia Dan-Cohen. *A Machine to Make a Future: Biotech Chronicles*. Princeton UP, 2005.

Ragan, Mark A., and Robert G. Beiko. "Lateral Genetic Transfer: Open Issues." *Philosophical Transactions of the Royal Society B: Biological Sciences*, vol. 364, no. 1527, Aug. 2009, pp. 2241–51, https://doi.org/10.1098/rstb.2009.0031.

Rai, Arti, and James Boyle. "Synthetic Biology: Caught between Property Rights, the Public Domain, and the Commons." *PLOS Biology*, vol. 5, no. 3, Mar. 2007, https://doi.org/10.1371/journal.pbio.0050058.

Raoult, Didier. "The Post-Darwinist Rhizome of Life." *The Lancet*, vol. 375, no. 9709, Jan. 2010, pp. 104–05, https://doi.org/10.1016/S0140-6736(09)61958-9.

Raoult, Didier, and Patrick Forterre. "Redefining Viruses: Lessons from Mimivirus." *Nature Reviews Microbiology*, vol. 6, no. 4, Apr. 2008, pp. 315–19, https://doi.org/10.1038/nrmicro1858.

Ray, Thomas S. "An Evolutionary Approach to Synthetic Biology: Zen and the Art of Creating Life." *Artificial Life: An Overview*, edited by Christopher G. Langton, MIT Press, 1997, pp. 179–210.

Reich, Karl A. "The Search for Essential Genes." *Research in Microbiology*, vol. 151, no. 5, June 2000, pp. 319–24, https://doi.org/10.1016/S0923-2508(00)00153-4.

Rich, Nathaniel. "The Mammoth Cometh." *The New York Times*, 27 Feb. 2014. *NYTimes.com*, https://www.nytimes.com/2014/03/02/magazine/the-mammoth-cometh.html.

Roberts, Dorothy E. *Fatal Invention: How Science, Politics, and Big Business Re-Create Race in the Twenty-First Century*. New Press, 2011.

Rodrigues, Rodrigo Araújo Lima, et al. "Translating the Language of Giants: Translation-Related Genes as a Major Contribution of Giant Viruses to the Virosphere." *Archives of Virology*, vol. 165, no. 6, June 2020, pp. 1267–78, https://doi.org/10.1007/s00705-020-04626-2.

Roosth, Sophia. *Synthetic: How Life Got Made*. U of Chicago P, 2017.

Rose, Nikolas. "The Human Sciences in a Biological Age." *Theory, Culture & Society*, vol. 30, no. 1, Jan. 2013, pp. 3–34, https://doi.org/10.1177/0263276412456569.

———. *The Politics of Life Itself: Biomedicine, Power, and Subjectivity in the Twenty-First Century*. Princeton UP, 2007.

Rosenthal, Adam. *Prosthetic Immortalities: Biology, Transhumanism, and the Search for Indefinite Life*. U of Minnesota P, 2024.

Roy, Deboleena. *Molecular Feminisms: Biology, Becomings, and Life in the Lab*. U of Washington P, 2018.

Ruska, H. "Über ein neues bei der bakteriophagen Lyse auftretendes Formelement." *Naturwissenschaften*, vol. 29, no. 24, June 1941, pp. 367–68, https://doi.org/10.1007/BF01479367.

Ryan, Frank. *Virusphere: From Common Colds to Ebola Epidemics — Why We Need the Viruses That Plague Us*. Prometheus, 2020.

Saldanha, Arun. "A Date with Destiny: Racial Capitalism and the Beginnings of the Anthropocene." *Environment and Planning D: Society and Space*, vol. 38, no. 1, Feb. 2020, pp. 12–34, https://doi.org/10.1177/0263775819871964.

Sandberg, Troy E., et al. "Adaptive Evolution of a Minimal Organism with a Synthetic Genome." *iScience*, vol. 26, no. 9, Sept. 2023, https://doi.org/10.1016/j.isci.2023.107500.

Sapp, Jan. *Genes, Germs and Medicine: The Life of Joshua Lederberg*. World Scientific, 2021.

———. *The New Foundations of Evolution: On the Tree of Life*. Oxford UP, 2009.

The SBL Study Bible. New Revised Standard Version, Updated Edition, HarperCollins, 2023.

Schaffer, Simon. "The Eighteenth Brumaire of Bruno Latour." *Studies in History and Philosophy of Science Part A*, vol. 22, no. 1, Mar. 1991, pp. 174–92, https://doi.org/10.1016/0039-3681(91)90020-S.

———. "Making Up Discovery." *Dimensions of Creativity*, edited by Margaret A. Boden, MIT Press, 1994, pp. 13–51, https://doi.org/10.7551/mitpress/2437.003.0004.

———. "Scientific Discoveries and the End of Natural Philosophy." *Social Studies of Science*, vol. 16, no. 3, Aug. 1986, pp. 387–420, https://doi.org/10.1177/030631286016003001.

Schrader, Astrid. "Microbial Suicide: Towards a Less Anthropocentric Ontology of Life and Death." *Body & Society*, vol. 23, no. 3, Sept. 2017, pp. 48–74, https://doi.org/10.1177/1357034X17716523.

———. "Responding to *Pfiesteria piscicida* (the Fish Killer): Phantomatic Ontologies, Indeterminacy, and Responsibility in Toxic Microbiology." *Social Studies of Science*, vol. 40, no. 2, Apr. 2010, pp. 275–306, https://doi.org/10.1177/0306312709344902.

Schrödinger, Erwin. *What Is Life?* Cambridge UP, 2016.

Scown, Corinne D., and Jay D. Keasling. "Sustainable Manufacturing with Synthetic Biology." *Nature Biotechnology*, vol. 40, no. 3, Mar. 2022, pp. 304–07, https://doi.org/10.1038/s41587-022-01248-8.

Senatore, Mauro. *Germs of Death: The Problem of Genesis in Jacques Derrida*. State U of New York P, 2019.

Serres, Michel. *The Parasite*. Translated by L. R. Schehr, U of Minnesota P, 2013.

Serres, Michel, and Bruno Latour. *Michelle Serres with Bruno Latour: Conversations on Science, Culture, and Time*. Translated by Roxanne Lapidus, U of Michigan P, 1995.

Shapin, Steven. "I'm a Surfer." *London Review of Books*, vol. 30, no. 6, Mar. 2008, https://www.lrb.co.uk/the-paper/v30/n06/steven-shapin/i-m-a-surfer.

Shapiro, Beth. *How to Clone a Mammoth: The Science of De-Extinction*. Princeton Science Library, 2020.

Shatilovich, A. V., et al. "Viable Nematodes from Late Pleistocene Permafrost of the Kolyma River Lowland." *Doklady Biological Sciences*, vol. 480, no. 1, May 2018, pp. 100–02, https://doi.org/10.1134/S0012496618030079.

Shreeve, James. *The Genome War: How Craig Venter Tried to Capture the Code of Life and Save the World*. Alfred A. Knopf, 2004.

Simons, Massimiliano. "Synthetic Biology as a Technoscience: The Case of Minimal Genomes and Essential Genes." *Studies in History and Philosophy of Science Part A*, vol. 85, Feb. 2021, pp. 127–36, https://doi.org/10.1016/j.shpsa.2020.09.012.

Staley, James T., and Allan Konopka. "Measurement of In Situ Activities of Nonphotosynthetic Microorganisms in Aquatic and Terrestrial Habitats."

Annual Review of Microbiology, vol. 39, no. 1, Oct. 1985, pp. 321–46, https://doi.org/10.1146/annurev.mi.39.100185.001541.

Stanton, Morgan M., et al. "Prospects for the Use of Induced Pluripotent Stem Cells in Animal Conservation and Environmental Protection." *Stem Cells Translational Medicine*, vol. 8, no. 1, Jan. 2019, pp. 7–13, https://doi.org/10.1002/sctm.18-0047.

Stent, Gunther S. *Molecular Biology of Bacterial Viruses*. W. H. Freeman, 1963.

Summers, William C. *Félix d'Herelle and the Origins of Molecular Biology*. Yale UP, 1999. eBook Academic Collection.

Suttle, Curtis A. "Marine Viruses — Major Players in the Global Ecosystem." *Nature Reviews Microbiology*, vol. 5, no. 10, Oct. 2007, pp. 801–12, https://doi.org/10.1038/nrmicro1750.

Synthetic Genomics Inc. "Synthetic Genomics Inc. and J. Craig Venter Institute Form New Company, Synthetic Genomics Vaccines Inc. (SGVI), to Develop Next Generation Vaccines." *PRNewswire*, 7 Oct. 2010, https://www.prnewswire.com/news-releases/synthetic-genomics-inc-and-j-craig-venter-institute-form-new-company-synthetic-genomics-vaccines-inc-sgvi-to-develop-next-generation-vaccines-104467694.html.

———. "Synthetic Genomics, Inc. Announces Agreement with New England Biolabs to Launch AssemblyTM Master Mix Product for Synthetic and Molecular Biology Applications." *PRNewswire*, 7 Feb. 2012, https://www.prnewswire.com/news-releases/synthetic-genomics-inc-announces-agreement-with-new-england-biolabs-to-launch-gibson-assembly-master-mix-product-for-synthetic-and-molecular-biology-applications-138884054.html.

———. "Synthetic Genomics Inc. Applauds the Venter Institute's Work in Creating the First Synthetic Bacterial Cell." *PRNewswire*, 20 May 2010, https://www.prnewswire.com/news-releases/synthetic-genomics-inc-applauds-the-venter-institutes-work-in-creating-the-first-synthetic-bacterial-cell-94478544.html.

Takahashi, Kazutoshi, and Shinya Yamanaka. "A Decade of Transcription Factor-Mediated Reprogramming to Pluripotency." *Nature Reviews Molecular Cell Biology*, vol. 17, no. 3, Mar. 2016, pp. 183–93, https://doi.org/10.1038/nrm.2016.8.

Tan, Xiao, et al. "Synthetic Biology in the Clinic: Engineering Vaccines, Diagnostics, and Therapeutics." *Cell*, vol. 184, no. 4, Feb. 2021, pp. 881–98, https://doi.org/10.1016/j.cell.2021.01.017.

Telesis Bio, Inc. "Home." *Telesis Bio*, 16 Dec. 2022, https://telesisbio.com/.

Thrasher, Steven W. *The Viral Underclass: The Human Toll When Inequality and Disease Collide*. Celadon Books, 2022.

Timár, Eszter. "Derrida's Error and Immunology." *Oxford Literary Review*, vol. 39, no. 1, July 2017, pp. 65–81, https://doi.org/10.3366/olr.2017.0210.

Tsing, Anna Lowenhaupt. *The Mushroom at the End of the World: On the Possibility of Life in Capitalist Ruins*. Princeton UP, 2015.

Twort, F. W. "An Investigation on the Nature of Ultra-Microscopic Viruses." *The Lancet*, vol. 186, no. 4814, Dec. 1915, pp. 1241–43, https://doi.org/10.1016/S0140-6736(01)20383-3.

USPTO. "Manual of Patent Examining Procedure." *The United States Patent and Trademark Office*, July 2022, https://www.uspto.gov/web/offices/pac/mpep/s2106.html.

Van Etten, James L., et al. "DNA Viruses: The Really Big Ones (Giruses)." *Annual Review of Microbiology*, vol. 64, 2010, pp. 83–99, https://doi.org/10.1146/annurev.micro.112408.134338. PubMed.

Van Helvoort, Ton. "The Construction of Bacteriophage as Bacterial Virus: Linking Endogenous and Exogenous Thought Styles." *Journal of the History of Biology*, vol. 27, no. 1, Mar. 1994, pp. 91–139, https://doi.org/10.1007/BF01058628.

Various. "What's in a Name?" *Nature Biotechnology*, vol. 27, no. 12, Dec. 2009, pp. 1071–73, https://doi.org/10.1038/nbt1209-1071.

Vartoukian, Sonia R., et al. "Strategies for Culture of 'Unculturable' Bacteria." *FEMS Microbiology Letters*, vol. 309, no. 1, 2010, pp. 1–7, https://doi.org/10.1111/j.1574-6968.2010.02000.x.

Venter, J. Craig. "A Conversation with J. Craig Venter, PhD." *Industrial Biotechnology*, vol. 12, no. 3, June 2016, pp. 134–36, https://doi.org/10.1089/ind.2016.29035.jcv.

———. *Life at the Speed of Light: From the Double Helix to the Dawn of Digital Life*. Little, Brown Book Group, 2013.

Venter, J. Craig, et al. Synthetic genomes. EP 1968994 B1, European Patent Office, 3 July 2013. *Google Patents,* https://patents.google.com/patent/EP1968994B1/.

Ventola, C. "The Antibiotic Resistance Crisis: Part 1: Causes and Threats." *P T*, vol. 40, no. 4, Apr. 2015, pp. 277–83.

Vergès, Françoise. "Racial Capitalocene." *Futures of Black Radicalism*, edited by Gaye Theresa Johnson and Alex Lubin, Verso Books, 2017.

Vierkant, Artie, and Beatrice Adler-Bolton. "The Year the Pandemic 'Ended' (Part 1)." *The New Inquiry*, 21 Dec. 2022, https://thenewinquiry.com/the-year-the-pandemic-ended-part-1/.

———. "The Year the Pandemic 'Ended' (Part II)." *The New Inquiry*, 22 Dec. 2022, https://thenewinquiry.com/the-year-the-pandemic-ended-part-ii/.

———. "The Year the Pandemic 'Ended' (Part III)." *The New Inquiry*, 23 Dec. 2022, https://thenewinquiry.com/the-year-the-pandemic-ended-part-iii/.

Villarreal, Luis P. *Viruses and the Evolution of Life*. ASM Press, 2005.

Vincent, James. "Scientists Rename Human Genes to Stop Microsoft Excel from Misreading Them as Dates." *The Verge*, 6 Aug. 2020, https://www.theverge.com/2020/8/6/21355674/human-genes-rename-microsoft-excel-misreading-dates.

Vitale, Francesco. *Biodeconstruction: Jacques Derrida and the Life Sciences*. Translated by Mauro Senatore, State U of New York P, 2018.

———. "Microphysics of Sex: Sexual Differences between Biology and Deconstruction." *Parallax*, vol. 25, no. 1, Jan. 2019, pp. 92–109, https://doi.org/10.1080/13534645.2019.1570608.

Wagner, Günter P. "The Biological Homology Concept." *Annual Review of Ecology and Systematics*, vol. 20, no. 1, Nov. 1989, pp. 51–69, https://doi.org/10.1146/annurev.es.20.110189.000411.

———. "The Developmental Genetics of Homology." *Nature Reviews Genetics*, vol. 8, no. 6, June 2007, pp. 473–79, https://doi.org/10.1038/nrg2099.

———. *Homology, Genes, and Evolutionary Innovation*. Princeton UP, 2014.

Wakayama, T., et al. "Full-Term Development of Mice from Enucleated Oocytes Injected with Cumulus Cell Nuclei." *Nature*, vol. 394, no. 6691, July 1998, pp. 369–74, https://doi.org/10.1038/28615.

Waltham-Smith, Naomi. *Shattering Biopolitics: Militant Listening and the Sound of Life*. Fordham UP, 2021.

Waterson, A. P., and Lise Wilkinson. *An Introduction to the History of Virology*. Cambridge UP, 1978.

Weintraub, Karen. "20 Years after Dolly the Sheep Led the Way — Where Is Cloning Now?" *Scientific American*, 5 July 2016, https://www-scientificamerican-com.eu1.proxy.openathens.net/article/20-years-after-dolly-the-sheep-led-the-way-where-is-cloning-now/.

Wells, Jonathan. *The Myth of Junk DNA*. Discovery Institute Press, 2011.

West-Eberhard, Mary Jane. *Developmental Plasticity and Evolution*. Oxford UP, 2003.

Westervelt, Amy. "Big Oil Firms Touted Algae as Climate Solution. Now All Have Pulled Funding." *The Guardian*, 17 Mar. 2023. *The Guardian*, https://www.theguardian.com/environment/2023/mar/17/big-oil-algae-biofuel-funding-cut-exxonmobil.

Wilkins, John S. *Species: The Evolution of the Idea*. Second Edition, CRC Press, 2018.

Willyard, Cassandra. "New Human Gene Tally Reignites Debate." *Nature*, vol. 558, no. 7710, June 2018, pp. 354–55, https://doi.org/10.1038/d41586-018-05462-w.

Wilson, Elizabeth A. *Psychosomatic: Feminism and the Neurological Body*. Duke UP, 2004.

Wilson, Malcolm. "Analogy in Aristotle's Biology." *Ancient Philosophy*, vol. 17, no. 2, 1997, pp. 335–58, https://doi.org/10.5840/ancientphil199717238.

Wimmer, Eckard. "The Test-Tube Synthesis of a Chemical Called Poliovirus." *EMBO Reports*, vol. 7, no. S1, 2006, pp. S3–9, https://doi.org/10.1038/sj.embor.7400728.

Wimmer, Eckard, and Aniko V. Paul. "Synthetic Poliovirus and Other Designer Viruses: What Have We Learned from Them?" *Annual Review of Microbiology*, vol. 65, no. 1, 2011, pp. 583–609, https://doi.org/10.1146/annurev-micro-090110-102957.

Xavier, Joana C., et al. "Systems Biology Perspectives on Minimal and Simpler Cells." *Microbiology and Molecular Biology Reviews*, vol. 78, no. 3, 2014, pp. 487–509, https://doi.org/10.1128/MMBR.00050-13.

Yong, Ed. "Giant Virus Resurrected from 30,000-Year-Old Ice." *Nature*, Mar. 2014, https://doi.org/10.1038/nature.2014.14801.

———. "Giant Viruses Open Pandora's Box." *Nature*, July 2013, https://doi.org/10.1038/nature.2013.13410.

Zeeberg, Barry R., et al. "Mistaken Identifiers: Gene Name Errors Can Be Introduced Inadvertently When Using Excel in Bioinformatics." *BMC Bioinformatics*, vol. 5, no. 1, June 2004, p. 80, https://doi.org/10.1186/1471-2105-5-80.

Zhang, Joy, et al. *The Transnational Governance of Synthetic Biology: Scientific Uncertainty, Cross-Borderness and the "Art" of Governance*. BIOS, 2011.

Ziemann, Mark, et al. "Gene Name Errors Are Widespread in the Scientific Literature." *Genome Biology*, vol. 17, no. 1, Aug. 2016, p. 177, https://doi.org/10.1186/s13059-016-1044-7.

Index